THE MATHEMATICS OF HEREDITY

THE MATHEMATICS OF HEREDITY

Gustave Malécot
University of Lyon

Revised, edited, and translated by
Demetrios M. Yermanos
University of California, Riverside

W. H. Freeman and Company
San Francisco

Translated from the French edition, copyright © 1948
by Masson et Cie, Éditeurs.

Library of Congress Catalog Card Number 69-12603.
Standard Book Number: 7167-0678-4

Contents

Author's Preface   vii
Translator's Foreword   ix
Author's Preface to the French edition   xi
Preface to Probabilité et Hérédité   xiii

1

The Mendelian Lottery   1

1.1 Heredity and the Laws of Mendel   1
1.2. The Chromosomes   4
1.3. Resemblance Between Related Individuals   8

2

Correlation Between Relatives in an Isogamous Stationary Population   13

2.1. Probabilities of Genes and Genotypes   13
2.2. The Distribution of Factors in an Isogamous Population   16
2.3. Random Mendelian Variables in
      an Isogamous Stationary Population   18

2.4. Correlations Between Relatives Without Dominance   **22**

2.5. Correlations Between Unrelated Individuals
with Dominance   23

2.6. Correlations Between Any Individuals with Dominance   **29**

3

Evolution of a Mendelian Population   31

3.1. Influence of Population Size on Neutral Genes   31

3.2. Influence of Selection   41

3.3. Influence of Migration   64

3.4. Appendix: Discontinuous Migrations   77

Bibliography   85

Index   87

Author's Preface

Many papers since my 1948 book have presented numerous applications of the ideas sketched in it, particularly about coancestry and migration; therefore, in this revised, English edition, I have added a few explanatory footnotes, and some formulas about the decrease of coancestry with distance. For further information the reader may use the new references added to the original bibliography, or my recent book [16].

I am grateful to Professor D. M. Yermanos for his many suggestions and corrections in revising this text and for the care with which he has edited and translated it.

Lyon, 1968   G. MALÉCOT

Translator's Foreword

The need for an English translation of Professor Gustave Malécot's classic work, *The Mathematics of Heredity*, has been known for some time by students of population genetics interested in his approach to dealing with problems of population structure. The lack of such a translation has curtailed the dissemination of his ideas among English-speaking biologists. We are now increasingly concerned with population science, yet there are few books in this field. I hope that this revised, English edition of Professor Malécot's book will not only enrich the literature now available, but also help bring his work the recognition it deserves.

The Preface by Professor Newton Morton to *Probabilités et Hérédités*, published in 1966 by the Presses Universitaires de France, summarizes well some of the significant aspects of Professor Malécot's work, and I have included it here with the kind permission of both Professor Morton and the Presses Universitaires de France.

September 1969   D. M. YERMANOS

Author's Preface to the French Edition

The objective of this work is the application of probability theory to prove a number of classical formulas as well as a few unpublished ones pertaining to genetics and the mathematical theory of evolution. Instead of suggesting a unique approach, which would have seemed too abstract to the biologist, I have preferred to present various methods, each adapted to a concrete problem; once the fundamental concepts of mathematical genetics are thus simplified, the foundations will have been laid for experimentation, which is indispensable, and the way will be clear for eventual synthesis. I apologize for the imperfections of this first text, and I will accept with interest all remarks and criticism that anyone would care to make. In particular, I would welcome comments on whatever relates to the theory of migration, published here for the first time, and which must be matched with experimental data.

I express my gratitude to Professor G. Darmois and the Institute of Statistics in Paris for making this work possible. Also, I express my appreciation to Professor L. Blaringhem for his valuable encouragement and to Masson et Cie for the care with which they have published this book.

Lyon, 1948   G. MALÉCOT

Preface to
Probabilités et Hérédité

The probabilistic theory of genetic relationship and covariance developed by Malécot has been propagated by disciples in other countries, notably Crow in the United States, Yasuda and Kimura in Japan, and Falconer in Great Britain, and is now universally accepted. The application of his results for isolation by distance, begun by Lamotte with *Cepea* and continued by Yasuda in man, promises to reveal population structure and the forces that have acted on major genes.

Malécot's insight is the more remarkable because Fisher, Haldane, and Wright, the great figures of population genetics in the older generation, used correlation analysis and did not mind that the derivation of correlations from probabilities is far easier than the reverse passage. By mid-century a reaction was inevitable. Major genes for blood groups, serum proteins, and other polymorphisms, as well as lethals and detrimentals, have become the heart of population genetics, and for them correlation partitions are inappropriate. At the same time, the invalidity of models of population structure based on genetic "islands" and "neighborhoods" has become apparent.

From *Probabilités et Hérédité* by Gustave Malécot, Presses Universitaires de France, 1966.

A probabilistic approach was begun by Cotterman, who in 1941 set forth the conditional probabilities for many kinds of relationship. His work had little impact, however, largely because the material of his thesis was published in summary, but also because his formulation was designed for nearly panmictic human populations and did not reveal the full power of probability methods.

Malécot's thesis of 1939 followed the classical approach of Fisher and Wright. His book of 1948, however, contains in a brief 63 pages a profoundly original treatment of relationship, covariance, and population structure in terms of probability theory. Every derivation began with the genotypic probabilities for a single locus, and with astonishing clarity the most complicated properties of Mendelian populations were revealed. Malécot identified Wright's coefficient of inbreeding, one of the great unifying concepts of mathematical biology, as the probability that uniting gametes are identical by descent, and introduced the more general coefficient of kinship (*parente*) to measure relationship of individuals possibly separated in time, space, or by other barriers, from which mating pairs are not randomly drawn. He replaced Wright's bewildering diversity of inbreeding coefficients relative to different subpopulations by one absolute measure of isolation by distance, the relation between the mean coefficient of kinship or inbreeding and the marital distance between birth places of potential mates. This led Wright to reexamine his results and conclude that "neighborhood size," on which his analysis of population structure is based, is almost independent of Malécot's basic relation between consanguinity and distance. There seems little doubt that research on population structure in the foreseeable future will follow the direction set by Malecot.

His later work on population structure was mathematically difficult, and publications in the *Annales de l'Université de Lyon* did not receive immediate recognition. As recently as 1964 Kimura and Weiss rediscovered the formula for two-dimensional isolation by distance which had been published by Malécot (1959 and earlier), and believed that their result was new. This book is therefore doubly welcome, as an orderly presentation of principles and as vindication of the priority of a great French savant who has transformed population genetics.

Professor of Genetics   NEWTON E. MORTON
University of Hawaii
Honolulu, 1966

THE MATHEMATICS OF HEREDITY

Chapter 1

The Mendelian Lottery

1.1 HEREDITY AND THE LAWS OF MENDEL

Let us recall the laws of Mendel, taking the four o'clock (*Mirabilis jalapa*) as an example. If we cross a white-flowered plant with a red-flowered one, we obtain only pink-flowered plants. But if we cross pink-flowered plants among themselves, we obtain progeny 1/4 of which, on the average, have white flowers, 1/2 pink flowers and 1/4 red flowers. The traits of the grandparents reappear. This is the phenomenon of Mendelian disjunction, or segregation. It can be explained by postulating that flower color in the four o'clock is determined by a pair of hereditary units or factors, each of which can appear in one or the other of two states or genes, which we will designate by *A* or *a*. Thus, an individual could carry the pairs *AA*, and so have red flowers; *Aa*, and have pink flowers; or *aa*, and have white flowers. The three states in which the pair can appear are called genotypes or zygotes. *AA* and *aa* are the homozygotes; *Aa* is the heterozygote.

The outcome of the cross can be interpreted by the following mechanism. The pair of factors of each plant resulting from the cross, i.e., of each "offspring," is obtained by drawing, at random,

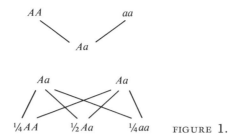

FIGURE 1.

one of the two factors of the father and one of the two factors of the mother (see Fig. 1). The cross of an AA with an aa gives only Aa (first generation), but the cross of Aa genotypes among themselves gives: AA with a probability of 1/4 (of drawing an A from both parents); Aa, 1/2; and aa, 1/4. This interpretation agrees well with the many observations of frequencies for individuals in the second generation.

The laws of Mendel explain remarkably well all the phenomena of heredity, and it can be said that, with a few rare exceptions, all heredity follows the Mendelian process. One has to admit, however, that genes can also act in a way different from the one just described.

(A) Let us consider the example given by Mendel of the cross between peas with round seeds and peas with wrinkled seeds. In the first generation, we obtain only peas with round seeds; when we cross these peas among themselves, 3/4 of their progeny have round seeds and 1/4 have wrinkled seeds. This result fits the previous scheme perfectly if we postulate that both AA and Aa have round seeds, and that only aa has wrinkled seeds.

In this case, the heterozygote Aa has the same external appearance as an AA homozygote, from which it cannot be distinguished except by the characteristics of its progeny; that is, we must distinguish the genotype, or hereditary constitution, from the phenotype, or external appearance. Here the three genotypes give only two phenotypes. The gene A is dominant over the gene a, i.e., a is recessive, and the heterozygote exhibits the same phenotype as the dominant

homozygote. Dominance can be incomplete, so that the heterozygote is closer to one of the homozygotes, but is nevertheless
distinct.

(B) Characteristics determined by several pairs of factors are
called multifactorial. For example, the shape of a rooster's comb
depends on three pairs of factors: the first pair with genes C (presence
of comb) dominant over c (rudimentary comb); the second pair R
(rose) dominant over r (single); the third D (double) dominant over d
(single). As a result of dominance, the genotypes $CCRRdd$, $CcRRdd$,
$CcRrdd$, and $CCRrdd$ produce the same phenotype, the rose comb,
but $CCrrdd$ and $Ccrrdd$ produce a single comb, and $ccrrDD$ and
$ccrrDd$ produce the pea comb (double rudimentary comb).

The study of crosses shows that segregation of different pairs
takes place independently. For example, crossing a breed of chickens
with a rose comb produced by the double heterozygote $CcRrdd$
with a breed having a pea comb of genotype $ccrrDD$ produces
progeny which all have the pair Dd, but which have either Cc or cc,
and either Rr or rr, each pair having a probability of $1/2$. Therefore,
since these two pairs undergo segregation independently, $1/4$ of the
genotypes of the progeny will be, on the average, $CcRrDd$, $1/4$ will
be $ccRrDd$, $1/4$ $CcrrDd$, and $1/4$ $ccrrDd$.

(C) In the three examples so far, the characteristics observed con-
stituted a discontinuous series. Karl Pearson, who was, with Frances
Galton, the founder of biometry, distinguished from this "alternative" heredity the "continuous" or "blending" heredity, as, for
example, that of stature or skin color in the human species. If one
observes enough children from a given couple, one finds that the
statures of the children are grouped around a mean value which
depends on the statures of the two parents, and conform to a bell-
shaped curve, with extreme deviations being rare but possible. It
seems that there is a blending of the parental characters, complicated
by fluctuations. In the same way, when two mulattoes marry, their
children can vary greatly in skin color; although most of the children

will have skin color more or less like that of the parents, from time to time a completely black or completely white genotype also occurs.* All these results are perfectly explained by Mendel's laws, on the basis that stature or skin color results from the accumulated effects of a large number, n, of Mendelian factors which disjoin independently. To illustrate the point, if, in each pair of factors, one of the two possible genes adds 1 mm to, and the other subtracts 1 mm from, the mean stature, and if one crosses two individuals in whom all pairs are heterozygous, $A_1a_1, A_2a_2, \ldots A_na_n$, in the offspring each pair of factors will have the probabilities 1/4, 1/2, and 1/4 of being in the states A_iA_i, A_ia_i, and a_ia_i, and of contributing to the stature 2 mm, 0 mm, and -2 mm, respectively, as in two independent random choices.

If the n different pairs are stochastically independent, the stature of a child will be nothing but the sum of gains and losses in $2n$ independent series of random choices. This sum, as we know, for an indefinitely large n, follows Gauss's law of probability, which fits the experimental bell-shaped curve. We shall see that the same is true whether one suppose dominance to be generally complete or generally incomplete in each pair or suppose different contributions for different pairs. This general scheme of multifactorial Mendelian heredity will be developed in detail in the second chapter, and will be shown to explain the results of biometry as discovered by Galton and Pearson [**5, 17, 18**].†

1.2 THE CHROMOSOMES

The physiological basis of Mendelian heredity was discovered in the small rods or *chromosomes* that are constituents of the nucleus of the reproductive cells or *gametes*. These chromosomes have a fixed number, n, in each species (23 in man, 4 in drosophila) and some-

---

* To be precise, one black and one white child will occur in every fourteen children, on the average. D. M. Y.

† Bold numbers in brackets refer to the literature cited at the end of the book.

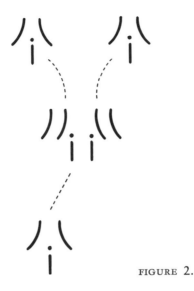

FIGURE 2.

times exhibit differences among themselves which make it possible to recognize in two different gametes the *homologous* chromosomes (see Fig. 2 for drosophila). When a paternal gamete unites with a maternal gamete, the fertilized egg has n pairs of homologous chromosomes. This chromosomal constitution persists in all the cells that the egg produces by division and, finally, in all the cells of the adult individual except the reproductive cells or gametes; the latter are produced by a division or *disjunction* which allows only one chromosome of each pair to be included in each gamete, this chromosome being taken at random from the two. Union of these gametes, at random, with the gametes of the other parent produces the individuals of the following generation.

The laws of Mendel are explained, therefore, by postulating that the two factors of a pair are carried by two homologous chromosomes. Two heterozygous parents, *Aa*, will each form gametes one half of which are *A* and one half *a*; this is *disjunction*. Random union of these gametes will produce offspring in the ratio 1/4 *AA*, 1/2 *Aa*, and 1/4 *aa*.

A difficulty appears, however; the different pairs of factors can be independent only if each pair is carried by a different pair of chromosomes, each of the latter being expected to disjoin independently. Factors carried by the same pair of chromosomes should be completely linked. However, given a genotype in which $\overline{AB}\,\overline{ab}$ represents two pairs of factors linked onto two chromosomes, if this genotype is crossed with a homozygote $\overline{ab}\,\overline{ab}$, the genotypes actually found in the progeny are *ABab* and *abab*, each with a frequency of $(1 - r)/2$, and *Abab* and *aBab*, each with a frequency of $r/2$, where r is, in general, a small positive number, but not zero as it would be if linkage were complete. These ratios can be explained only by postulating that the chromosomes break and exchange homologous sections before disjunction. The four types of gametes formed have the given frequencies because of this exchange, whose probability of occurring is r. This is the phenomenon of "crossing over." A study of this phenomenon leads to the assumption that each factor is localized at a specific point on the chromosome; this point is its *locus* (plural, *loci*). For two factors found on the same chromosome, it is evident that the farther apart they are located on the chromosome, the greater will be the probability r of their crossing over. This phenomenon made it possible to map the loci of the different factors for the four pairs of chromosomes of drosophila. We shall see in the following chapter that, because of crossing over, linkage of factors on the same chromosome does not prevent them, in the long run, from being as reshuffled as if they were independent.

Sex is determined by a pair of chromosomes, two X chromosomes for the female, an X and a Y chromosome for the male (except in lepidoptera and birds), called heterosomes; the other chromosomes are called autosomes. The factors carried on the heterosomes are called "sex-linked"; we know primarily those carried on the X chromosome. These other factors are never masked in males, but can be in females (as with daltonism and hemophilia).

Whatever the physiological process by which genes affect the development of individuals, it is to be expected that a single pair of genes can affect several characteristics. For example, the recessive gene for albinism produces, in the homozygous recessive *aa*, both white hair and red eyes (due to lack of pigment). On the other hand, a single characteristic, such as stature, can be influenced by many pairs of genes. Although, strictly speaking, each characteristic depends on the entire gene complement—on the total "genetic constitution"—each characteristic is in fact influenced appreciably by only one pair of genes, which made it possible for Mendel to deduce his laws.

We have so far been tacitly assuming that the genes occupying a specific locus can appear in only two different states, represented by *A* and *a*, which we call *allelic* genes. In reality, they can also have multiple states A', A'', A''', $\ldots A^{(n)}$, which we call *multiallelism*. There are *n* homozygotes and $_nC_2 = \dfrac{n(n-1)}{2}$ heterozygotes.* The origin of allelic genes is to be traced back to the phenomenon of mutation, which appears to be (at least according to present observations, since one cannot guess what took place in paleontological times) the only inheritable way in which living organisms can be modified and, therefore, the only one that affects the evolution of species. Mutation is an abrupt change affecting one of two homologous loci in an individual; this change is, therefore, transmitted to one half its gametes. Thus, in a population of homozygous *AA* individuals, which are indistinguishable in terms of this pair of genes, there may appear an *Aa* heterozygote, which, even though *A* is dominant, can be identified by its progeny. Mutation produced the new gene *a*, which is allelic to the old gene *A*. Repeated mutations affecting the same locus can continue to create the same gene *a* (recurrent mutation), or cause gene *a* to revert to *A* (reverse mutation), or create other alleles (multiallelism).

---

* For example, the four blood groups in man are determined by three alleles, A, B, and O, A and B being dominant over O, which give the four phenotypes *A* (genotype *AA* or *AO*), *B* (*BB* or *BO*), *AB* (universal recipient), and *O* (*OO*) (universal donor).

1.3 RESEMBLANCE BETWEEN RELATED INDIVIDUALS

Two individuals in a population are related if they have one or more common ancestors. If they do, their genetic difference must be smaller, on the average, than that between two individuals taken at random, because some of the genes of the first two are derived from the corresponding genes of the common ancestor. Disregarding mutations, these genes cannot be different, whereas they often could be in unrelated individuals.

For precision in terminology, let us distinguish between *factors*, *genes*, and *loci*. Let us call *genes* the different states in which each factor can appear without regard to the individual in which they are observed. Two genes corresponding to the same factor and observed either in the same individual or in two different individuals will be called identical or different, depending on whether they appear in the same state, for example, A, or in two allelic states, for example, A and a. However, two *loci* will be called identical only if they were derived by Mendelian descent from the same locus of the same common ancestor; otherwise they will be called different. Two identical loci are by necessity occupied by two identical genes, if there is no mutation, but two different loci can be occupied by either two identical or two different genes.

An individual, I, has two parents, four grandparents, . . . 2^n ancestors of order n. A locus of I has a probability of $1/2$ of being derived from the father, $1/2$ from the mother, $1/4$ from each of the grandparents, $1/2^n$ from each ancestor of order n, along a given chain of descent. An ancestor of I can be linked to it by several chains of descent; for example, J in Figure 3 is a multiple ancestor, and could even be an ancestor of different order in different chains.

We will designate as the *coefficient of coancestry*, f_{IL}, of two individuals I and L the probability that two homologous loci, one from I and the other from L, are identical, i.e., are descended from the same locus. The complementary probability, $1 - f_{IL}$, is the prob-

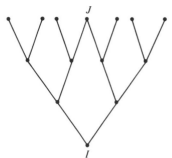

FIGURE 3.

ability that these two loci come from unrelated ancestors, i.e., that they are stochastically independent, since knowing the gene which occupies one locus does not provide any information about the gene which occupies the other; these two genes can be identical or different, but their probabilities are independent.

We shall designate as the *coefficient of inbreeding*, f_M, of an individual M the probability that its two homologous loci are identical. Since one locus is derived from its father and the other from its mother, f_M is nothing but the coefficient of coancestry of its two parents.

Let us evaluate the coefficient of coancestry, f_{IL}, of two individuals, I and L. It differs from zero only where I and L have one or more common ancestors, J_1, J_2, J_3, etc., which we will assume they have. Let us suppose at first that there is only one common ancestor, J, of order n for I and of order p for L along two distinct chains of descent, which together constitute a chain of coancestry linking I and L.

The probability that one locus of I and one homologous locus of L are both derived from J is $(1/2)^{n+p}$. If they are both derived from J, they have a probability of $1/2$ of being derived from the same locus of J and a probability of $1/2$ of being derived from different loci; if they are from different loci, the probability that they will be identical is f_J. From this, $f_{IL} = (1/2)^{n+p}(1 + f_J)/2$. In particular, the coefficient of coancestry of an individual and an ancestor of order n is given by letting $p = 0$; the coefficient of coancestry of an individual with itself is given by letting $n = p = 0$.

Let us now consider the general case, in which I and L are connected by any number of chains of coancestry, each chain being the combination of two chains of descent leading from I and from L to a common ancestor J_i and having no other common point except J_i; two chains of coancestry are considered distinct, even if they have links in common, provided they differ by at least one link. Since the transmission of identical loci along a specified chain of relationship excludes their transmission along any other, the principle of total probability gives

$$f_{IL} = \Sigma(1/2)^{n_i+p_i}(1 + fJ_i)/2.$$

The sum Σ extends over all distinct chains of relationship connecting I and L; the ith chain has $n_i + p_i$ links ascending from I and from L to the common ancestor J_i, whose coefficient of inbreeding is f_{J_i}.

For example, if we assume that all chains of relationship are shown in Figure 4 and that the ancestors A and B are not related, there are the following distinct chains and respective contributions to the coefficient f_{GF}: $GCF = 1/8$; $GEF = 5/32$; $GCAEF = 1/32$;

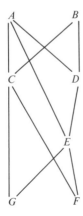

FIGURE 4.

$FCAEG = 1/32; GCADEF = 1/64; FCADEG = 1/64; GCBDEF =$
$1/64; FCBDEG = 1/64.$ Therefore, $f_{GF} = 13/32.$

Let us suppose now that the loci considered may have an average rate of mutation, u, per generation. The probability that a locus of an individual reproduces without modification the parental locus from which it was derived is $1 - u$; therefore, the probability that this locus may be transmitted without modification along a specified chain of descent having n links is $[(1 - u)/2]^n$. The coefficient of coancestry of two related individuals then becomes

$$\Sigma \left(\frac{1 - u}{2}\right)^{n_i + p_i} \frac{1 + f_{A_i}}{2}.$$

The correction thus introduced is insignificant for close relatives, since u is extremely small; it becomes important, as we shall see, only when very distant ancestors are involved.

Chapter 2

Correlations Between Relatives in an Isogamous Stationary Population

2.1 PROBABILITIES OF GENES AND GENOTYPES

Let us classify the individuals of a population F according to the states of a specified pair of factors. Let us first suppose that there are only two alleles, A and a, and, therefore, the three genotypes AA, Aa, aa, with the respective frequencies P, $2Q$, R, where $P + 2Q + R = 1$. Let us define the frequencies of genes A and a as the quantities p and q, where $p = P + Q$, $q = Q + R$, and $p + q = 1$.

These quantities are the probabilities that a gene taken at random from any individual in population F is in state A or a, respectively. In each individual I of the population F, each of the two homologous loci will be occupied by gene A or a with the probability p or q. However, there will generally be a relationship between the probabilities of these two loci, that is, a correlation between the states of these two loci, because knowing which gene occupies one of these loci affects the probabilities for the other locus. In fact, the two parents of the preceding generation from which these loci are descended could have been selected according to their relationship (consanguinity) or according to their resemblance (assortative mating), or they could have left only selected descendants because of differential fecundity; if so, any information on the genotype of one parent modifies the probabilities for the other. In this chapter we shall deal with the following two cases.

(A) The parents mate at random; the probability of finding a mate is the same for all individuals; and fecundity is the same for all couples. This is "random mating," panmixia. In this case, knowing the gene which occupies one of two loci of *I* gives us no information about the other; the states of these two loci are stochastically independent. Therefore, *I* may have one of the three genotypes *AA, Aa, aa*, with probabilities p^2, $2pq$, q^2. If the population is large, the observed frequencies *P*, $2Q$, *R*, must be close to these quantities. To prove this, it is sufficient to show that $Q^2 - PR$ is close to zero (Hardy's law), because we can set $P = p^2 + \lambda$, $2Q = 2pq - 2\mu$, $R = q^2 + \nu$, and since we have set $P + Q = p$, $Q + R = q$, and $p + q = 1$, we have $\lambda = \mu = \nu$; therefore,

$$Q^2 - PR = (pq - \lambda)^2 - (p^2 + \lambda)(q^2 + \lambda) = -\lambda,$$

which equals 0 only when $\lambda = 0$. Natural populations actually exist in which Hardy's law is confirmed, e.g., the population of coleoptera *Dermestes vulpinus* observed by Philip [19] (the pair of factors studied determines wing color). We shall see that there are such populations in the human species, too, for blood groups.

(B) The parents mate according to their consanguinity without considering their genotypes or resemblance; the probability of finding a mate is the same for all individuals; and all couples have the same fecundity. This is pure consanguinity or isogamy. Therefore, a locus in any individual, whether derived from a consanguineous cross or not, has always the same probabilities, *p* and *q*, of carrying the genes *A* or *a*; furthermore, for any individual *I* whose coefficient of inbreeding $f_I = f$, is known, the two homologous loci have, as we have seen, the probability *f* of being identical and the probability $1 - f$ of being stochastically independent; therefore, the three genotypes will occur with the probabilities, $fp + (1 - f)p^2 = p^2 + fpq$, $2(1 - f)pq$, and $fq + (1 - f)q^2 = q^2 + fpq$. (For example, to have the first genotype, the two loci should be identical and one of them should be A, or they should be independent and both of them should be A).

Consanguinity, therefore, causes an appreciable increase in the probability of homozygotes and a decrease in the probability of heterozygotes. This fact explains the danger of marriages between related persons; latent defects in the human species are generally determined by rare recessive genes, and appear only in homozygous recessives aa. If q is the frequency, presumably low, of a defective gene a, the probability that an individual I carries the defect, i.e., that it is of the genotype aa, will be equal to q^2 (which is extremely low) if the parents of I are not related; but this probability increases to $q^2 + fpq \sim fq$ if f_I is rather high. For example, a defect brought about by a gene with frequency $q = 10^{-4}$ will appear with the probability 10^{-8} in an offspring without inbreeding, but with the probability $10^{-4}/16$ in an offspring of first cousins ($f = 1/16$).* The danger is doubled for double first cousins ($f = 1/8$). It is thus unreasonable to tolerate marriage between double first cousins and between uncle and niece, and to forbid marriage between half-sibs, which presents exactly the same danger ($f = 1/8$) [6].

Let us consider now the more general case of multiallelism. Suppose that the allelic genes A_i have the frequencies p_i ($\Sigma p_i = 1$).

(1) With random mating, the probabilities of different genotypes are p_i^2 for an A_iA_i homozygote, $2p_ip_j$ for an A_iA_j heterozygote, these probabilities being coefficients in the expansion of $(\Sigma p_it_i)^2$. These formulas approximate well the frequencies of blood groups in a homogeneous population ($p^2 + 2pr$, $q^2 + 2qr$, $2pq$, r^2).

(2) In the more general case of isogamy, these probabilities, for an individual with coefficient of inbreeding f, are, respectively,

---

* In effect, it has been found that $1/2$ of the cases of Friedreich's ataxia, as well as $1/3$ of the cases of albinism, are derived from marriages between relatives.

$$fp_i + (1 - f)p_i^2 = p_i^2 + fp_i(1 - p_i) \quad \text{and} \quad 2(1 - f)p_ip_j,$$

these being coefficients in the expansion of $f\Sigma p_it_i^2 + (1 - f)(\Sigma p_it_i)^2$.

2.2 THE DISTRIBUTION OF FACTORS IN AN ISOGAMOUS POPULATION

Let us call "isogamous" a population, F, derived from parents chosen either at random or because of their coancestry (but excluding all homogamy), and in which all pairs have the same fecundity. Assume that the proportion of couples having a coefficient of coancestry f_i is w_i (the proportion w_0 corresponds to random mating, with $f_0 = 0$); w_i is, therefore, the frequency of individuals in the population with inbreeding coefficient f_i, and $\Sigma_i w_i = 1$. We have seen that the probabilities of the alleles A and a (assuming only two of them for simplicity) are the same among these individuals as in the total population, e.g., p and q.

The probabilities of the three genotypes in the entire population, and, accordingly, their frequencies, P, $2Q$, R, if the population is large, are

$$\Sigma w_ip(p + f_iq), \quad \Sigma 2w_ipq(1 - f_i), \quad \Sigma w_iq(q + f_ip),$$

which can also be written as

$$p(p + \alpha q), \quad 2pq(1 - \alpha), \quad q(q + \alpha p),$$

setting $\alpha = \Sigma w_if_i$; α is the mean inbreeding coefficient of the population, the mean of the coefficients of its individuals. It had been introduced *a priori* by Bernstein [2] to measure deviations from panmixia. His approximate evaluation has been tested on some human populations with the help of state census data on consanguineous marriages. In general, this coefficient is small: in a rural Austrian population, Reutlinger found α to be 0.6 per cent; in a Jewish population Orel found α to be a little over 1 per cent. These estimates, however, are probably much below the actual coefficient, because the distant relationships, which are overlooked, play as important a role as the close ones.

After considering the segregation of one pair of factors, let us now study the simultaneous segregation in the population F of two pairs of factors occupied by genes having the states A_i and B_j with probabilities p_i and x_j, respectively.

An individual I taken at random in F results from the union of two gametes from the preceding generation, F'. Let us call P'_{ij} the probability that any gamete Γ' coming from generation F' has in its chromosomes the genes A_i and B_j, and P_{ij} the probability that the same will be true for any gamete Γ from F, i.e., for a gamete produced by I; and let us find the relation between P_{ij} and P'_{ij}. When the gamete Γ produced by I has the genes A_i and B_j, either both are derived from the same gamete Γ' or each came from one of the two gametes Γ' which made up I. These two possibilities each have the probability $1/2$, if the two genes are found on two different chromosomes, because of independent segregation; but they have the probabilities $1 - r$ and r if the two genes are located on the same chromosome, because of "crossing over." The first possibility will be included with the second when $r = 1/2$. Then we have: $P_{ij} = (1 - r)P'_{ij} + r\pi_{ij}$, where π_{ij} is the probability that, in generation F, a gamete carrying A_i may unite with a gamete carrying B_j.

Different pairs, therefore, are not generally stochastically independent, since their distribution depends on the distribution in preceding generations, i.e., on an initial distribution which might be arbitrary. It will be shown, however, that there is an "asymptotic independence" under the following hypotheses.

(1) The population considered is very large, so that frequencies and probabilities in each generation are essentially equal.

(2) The population is isogamous, so that, as we have seen, no gene is favored; therefore, in each generation, the gene probabilities will remain equal to their frequencies in the preceding generation. As a result, the frequencies p_i will remain constant over generations. These will be the characteristic constants of the population and of

the system of alleles considered. From these, one can derive the probabilities of the three genotypes for individuals with coefficient of inbreeding f, or their frequencies if they are sufficiently numerous.

(3) The mating system adopted, although it implies a relationship between the two gametes that unite, leaves their probabilities of carrying different genes independent. This consequence, evident for panmixia, is not always valid in crosses between relatives, e.g., when the population is divided into groups between which crosses are impossible. It can be shown, for example, that it applies to brother-sister matings, if all individuals in each generation are brothers and sisters of one family; if not, the population would be distributed into several groups, and differences between genes existing in these groups would continue to exist indefinitely. Let us assume, therefore, that the mating system chosen is such that it leaves independent the probabilities of one uniting gamete carrying gene A_i, the other gene B_j. Then the probability π_{ij} of the union of a gamete carrying A_i with a gamete carrying B_j will be constant and equal to $p_i x_j$. The above recurrence equation may be written as

$$P_{ij} - p_i x_j = (1 - r)(P'_{ij} - p_i x_j).$$

If, therefore, the P_{ij} of one generation is equal to $p_i x_j$, it will always remain equal in the following generation; we say then that the population is stationary, and we note that the genes of the different pairs are stochastically independent. In a nonstationary population, $P'_{ij} - p_i x_j \longrightarrow 0$ as $(1 - r)^n \longrightarrow 0$ when the number, n, of generations tends to infinity, and the population tends to become stationary; let us assume in the remainder of the chapter that this state of equilibrium has been attained, and, in particular, that there is stochastic independence of the different factors.

2.3 RANDOM MENDELIAN VARIABLES IN AN ISOGAMOUS STATIONARY POPULATION

Let us consider a specific trait, e.g., stature, of the individuals that make up the population; this trait can be either quantitative and measurable, or qualitative and arbitrarily assigned to values on a numerical scale. Call y the numerical value thus attributed to the trait in each individual. For an individual I taken at random from the population, y is a random variable. We shall regard y as being the sum of a random variable, x, which represents the influence of the genetic constitution of I on the trait considered, and of another random variable, z, which represents the influence of chance and environment on the development of this trait, z being stochastically independent of x. Consider x the sum of contributions made to the trait by a certain number of pairs of factors. For example, the contribution $\mathcal{X}$ of one of its pairs will be equal to i, j, or k, depending on whether this pair has the state AA, Aa, or aa, whose probabilities are $p^2 + fpq$, $2(1 - f)pq$, and $q(q + fp)$, respectively, where p and q are the frequencies of A and a, and f is the inbreeding coefficient of I. $\mathcal{X}$ will be called the genotypic random variable associated with the trait and with the pair of factors considered.[*] If one of the alleles has complete dominance, $j = i$ or $j = k$. If there is no dominance, i.e., when the heterozygote is exactly intermediate between the two homozygotes, $j = (i + k)/2$, or one can let $i = 2t, j = s + t$, and $k = 2s$; one can readily verify that the three-valued random variable $\mathcal{X}$ is the sum of the two-valued random variables H and H', each of which has the value t or s with probabilities p or q, and which have the probability f of being identical and the probability $(1 - f)$ of being independent. H and H', which represent the respective states of the two loci of the pair, will be referred to as genic random variables. If there is dominance (complete or incomplete), we can still keep the random variables H and H' by taking appropriate values for s and t, and letting $\mathcal{X} = H + H' + d$, the component of dominance, d, being equal to $i - 2t, j - s - t$, or $k - 2s$, according to whether $H + H'$ is equal to $2t, s + t$, or $2s$ (the most

---

[*] Unless otherwise specified, the sampling unit on which this random variable depends is an individual taken at random from among those having inbreeding coefficient f.

convenient values for s and t will be discussed later). We have, therefore:

$$y = x + z = \Sigma \mathfrak{X} + z = \Sigma(H + H' + d) + z,$$

with Σ designating a sum over all pairs of factors influencing the trait under consideration (for a monofactorial trait, Σ covers only one term). Since the population under study is assumed to be stationary, the different terms of Σ are, as we have seen, independent random variables; z is also assumed to be independent.

To simplify matters, we shall assume, henceforth, that each of these random variables is given its mean value in the population (or in a specified subgroup of this population) as an origin. This assumption is not restrictive as long as we agree to stipulate that, in measuring the characteristic, we take its expectation as equal to 0, which is approximately the same as subtracting the general mean in the population (or in the subgroup) if it contains a large number of individuals. If we symbolize the expectation of a random variable by $\mathfrak{M}$, the stipulation which we made will be expressed by $\mathfrak{M}(\mathfrak{X}) = 0$, $\mathfrak{M}(x) = 0$, $\mathfrak{M}(z) = 0$, and $\mathfrak{M}(y) = 0$, and, by selecting appropriate values of s and t, $\mathfrak{M}(H) = 0$, $\mathfrak{M}(d) = 0$.

The variance of trait y (i.e., the square of its standard deviation) in the population (or in the subgroup) because of independence will be:

$$\mathfrak{M}(y^2) = \mathfrak{M}(x^2) + \mathfrak{M}(z^2) = \Sigma\mathfrak{M}(\mathfrak{X}^2) + \mathfrak{M}(z^2),$$

which we write as

$$\sigma_y^2 = \sigma_x^2 + \sigma_z^2,$$

σ being the standard deviation.

All these formulas are also valid for multiallelism.

The fact that a population is stationary imposes the condition that variance remain constant over generations. Thus, the fact that variance is conserved, as we know by experience, may be considered confirmation of the Mendelian hypothesis of the inheritance of characteristics. The theories of "blending inheritance" that certain biometricians tended to accept would imply that, if the hereditary portion x of a trait in the two parents was equal to x_1 and x_2, it would be equal to $(x_1 + x_2)/2$ in the offspring, the remainder of variance being attributed to chance and to the environment. Given panmixia, and assuming that $\mathfrak{M}(x) = 0$, the variance of x in the entire progeny of a population, taking $\mathfrak{M}(x_1 x_2) = 0$, would be $\mathfrak{M}[(x_1 + x_2)/2]^2 = \mathfrak{M}(x^2)/2$, i.e., half the variability in the parents; thus, the genetic variance of x would tend rapidly toward zero after several generations. Finally, the only variance left would be caused either by chance and environment (but the experiments of Johannsen [9] on pure lines have shown that such variance is small for most traits) or by mutations, which would then have to be very frequent (but this conclusion contradicts our experience). "Blending inheritance" is thus inadmissible, and the Mendelian scheme, with indefinite disjunction of parental traits, is one of the simplest of those that have the conservation of hereditary variance as a consequence [4].

Let us now show how the Mendelian scheme leads to the same results as biometry. Assume, henceforth, that the trait studied is multifactorial and depends on a large number of genes, each making a contribution of the same order of magnitude. Therefore, x is the sum of a great many independent random variables, each of which is small in relation to the standard deviation, σ_x, of x; according to Liapounov's theorem, the probability of x follows Gauss's law, $(1/\sqrt{2\pi\sigma_x}) \exp(-x^2/2\sigma_x^2) \, dx$. If the effects of chance and environment on development come from multiple and independent sources, z and y will also be almost Gaussian, which result agrees with the observations of Galton and Pearson on stature.

Let us measure the trait y for two related individuals, I_1 and I_2, and let y_1 and y_2 be the two respective values. It can be shown that the probability of the sum of the two random variables y_1 and y_2 follows closely Gauss's law and, therefore, can be expressed by its coefficient of correlation. The experimental determination of this coefficient for a large number of pairs of individuals with the same ancestry in a large population was made for different populations by Galton [5], Pearson [17, 18], and Snow [20]. We shall now consider its theoretical value.

2.4 CORRELATIONS BETWEEN RELATIVES WITHOUT DOMINANCE

Without dominance, we have the relationships

$$y = \Sigma\mathfrak{K} + z = \Sigma(H + H') + z$$

and

$$\mathfrak{M}(H + H') = 2\mathfrak{M}(H) = 2(pt + qs).$$

Therefore, the convention $\mathfrak{M}(\mathfrak{K}) = 0$ is equivalent to $\mathfrak{M}(H) = 0$, that is, $pt + qs = 0$.

Let y_1 and y_2 be the random variables representing the traits of two individuals, I_1 and I_2, with coefficient of coancestry f. Without dominance,

$$y_1 = \Sigma(H_1 + H_1') + z_1 = H_1 + H_1' + K_1 + K_1' + \ldots + z_1,$$

and

$$y_2 = \Sigma(H_2 + H_2') + z_2 = \ldots .$$

To find their coefficient of correlation, r, let us calculate the mean value of their product, which is reduced to

$$\mathfrak{M}(y_1 y_2) = \Sigma\mathfrak{M}(H_1 H_2 + H_1' H_2 + H_1 H_2' + H_1' H_2'),$$

since, because of independence, $\mathfrak{M}(z_1 H_2) = \mathfrak{M}(z_1)\mathfrak{M}(H_2) = 0$, and so on; $\mathfrak{M}(z_1 z_2) = 0$; and, if K and K' represent the genic random variables for any other pair, $\mathfrak{M}(K_1 H_2) = \mathfrak{M}(K_1)\mathfrak{M}(H_2) = 0$, and so on.

Furthermore, each term, such as $\mathfrak{M}(H_1 H_2)$, is calculated on the basis that the random variables H_1 and H_2 reflect the state of two homologous loci taken at random on I_1 and I_2; i.e., they have a probability f of being identical and $1 - f$ of being independent. From this,

$$\mathfrak{M}(H_1 H_2) = f\mathfrak{M}(H_1^2).$$

Thus, f is the correlation coefficient of H_1 and H_2. Therefore,

$$\mathfrak{M}(y_1 y_2) = 4f \Sigma \mathfrak{M}(H_1^2).$$

On the other hand, if f_1 and f_2 are the inbreeding coefficients of I_1 and I_2, we have

$$\mathfrak{M}(y_1^2) = \Sigma \mathfrak{M}(H_1 + H_1')^2 + \mathfrak{M}(z^2) = 2(1 + f_1) \Sigma \mathfrak{M}(H_1^2) + \mathfrak{M}(z^2),$$

since $\mathfrak{M}(H_1 H_1') = f_1 \mathfrak{M}(H_1^2)$. Therefore, the correlation coefficient sought is

$$r = \mathfrak{M}(y_1 y_2)/\sqrt{\mathfrak{M}(y_1^2)\mathfrak{M}(y_2^2)} = 2f/\sqrt{(1 + f_1 + \xi^2)(1 + f_2 + \xi^2)},$$

calling ξ^2 the ratio $\mathfrak{M}(z^2)/2\Sigma \mathfrak{M}(H_1^2)$.

For a trait determined by heredity only and for unrelated individuals, r reduces to $r_0 = 2f$, which will be called the fundamental correlation. This gives the familiar coefficients: $1/2$ for parent and offspring and for full sibs; $1/4$ for half-sibs, or for grandfather and grandson, or for uncle and nephew, or for double first cousins; $1/8$ for first cousins; and so on. But any coancestry between the individuals compared, and any effects of environment upon them, will make f_1, f_2, and ξ^2 unequal to zero, and will reduce the fundamental correlation.

2.5 CORRELATIONS BETWEEN UNRELATED INDIVIDUALS WITH DOMINANCE

Given that the probabilities of genes A and a are p and q, the probabilities of the three genotypes in the population will be p^2, $2pq$, and q^2. Let us still consider that the random variables $\mathfrak{X}$ have origins such that $\mathfrak{M}(\mathfrak{X}^2) = p^2 i + 2pq j + q^2 k = 0$; d takes the values $i - 2t$, $j - s - t$, and $k - 2s$, with probabilities p^2, $2pq$, q^2; t and s are the values that each of the random variables H and H' may take (values which, so far, are arbitrary). Along with Fisher [3], let us choose values which minimize $\mathfrak{M}(d^2)$; we obtain

$$p(i - 2t) + q(j - s - t) = 0,$$
$$p(j - s - t) + q(k - 2s) = 0, \tag{2.5.1}$$

by setting the partial derivatives with respect to t and s equal to zero. We thus obtain the fixed values, $t = pi + qj$, $s = pj + qk$, which satisfy the equation $\mathfrak{M}(H) = 0$ (because $pt + qs = 0$). Therefore,

$$\mathfrak{M}(d) = \mathfrak{M}(\mathfrak{K}) - \mathfrak{M}(H) - \mathfrak{M}(H') = 0.$$

Furthermore, equations (2.1) indicate that the mean value of d is zero when the value of H (or of H') is fixed. If we set H equal to t, H' (which is independent of H because the individuals are not related) will take the values t or s with probabilities p or q, and d will take the values of $i - 2t$ or $j - s - t$, whose mean value is equal to zero in accordance with (2.5.1). It follows that $\mathfrak{M}(dH) = \mathfrak{M}(dH') = 0$. Thus

$$\mathfrak{M}(\mathfrak{K}^2) = \mathfrak{M}(H + H')^2 + \mathfrak{M}(d^2) = 2\mathfrak{M}(H^2) + \mathfrak{M}(d^2),$$
$$\mathfrak{M}(y^2) = \Sigma 2\mathfrak{M}(H^2) + \Sigma \mathfrak{M}(d^2) + \mathfrak{M}(z^2).$$

Let us take for two related individuals, I_1 and I_2, the values $y_1 = \Sigma(H_1 + H_1' + d_1) + z_1$, and $y_2 = \Sigma(H_2 + H_2' + d_2) + z_2$. By hypothesis, H_1 and H_1' are independent, as are H_2 and H_2'. H_1, therefore, cannot be positively correlated at the same time with both H_2 and H_2'. Let us suppose, for example, that H_1 is correlated with H_2 only; in that case H_1' can only be correlated with H_2'; let us designate the respective correlation coefficients by ϕ and ϕ'. We have

$$(y_1 y_2) = \Sigma[\mathfrak{M}(H_1 H_2) + \mathfrak{M}(H_1' H_2') + \mathfrak{M}(d_1 d_2) + \mathfrak{M}(d_1 H_2)$$
$$+ \mathfrak{M}(d_1 H_2') + \mathfrak{M}(d_2 H_1) + \mathfrak{M}(d_2 H_1')], \quad (2.5.2)$$

because terms such as $\mathfrak{M}(z_1 H_2)$, $\mathfrak{M}(z_1 d_2)$, or $M(z_1 z_2)$ are equal to zero, since the random variables they include are independent and have mean values equal to zero.

Furthermore, the last four terms of (2.5.2) are also equal to zero. If, for example, the value of H_2 is fixed, d_2 depends only on H_2'; therefore, it is independent of H_1, and $\mathfrak{M}(H_1 d_2) = \mathfrak{M}(H_1)M(d_2)$, but we know $\mathfrak{M}(d_2) = 0$. Therefore, the mean value of $H_1 d_2$, being equal to zero when H_2 is fixed, is also equal to zero for any value of H_2. The same is true for the other three terms. Thus

$$\mathfrak{M}(y_1 y_2) = \Sigma[\mathfrak{M}(H_1 H_2) + \mathfrak{M}(H_1' H_2') + \mathfrak{M}(d_1 d_2)]$$
$$= (\phi + \phi')\Sigma\mathfrak{M}(H^2) + \Sigma\mathfrak{M}(d_1 d_2),$$

and everything goes back to the computation of $\mathfrak{M}(d_1 d_2)$.

2.5.1 The Two Individuals Are Related by Only One of Their Loci

In this situation, only two of the four genic random variables are not independent, e.g., H_1 and H_2; let ϕ be their correlation coefficient, which depends on their degree of relationship. If we fix H_1, then d_1 depends only on H_1' and becomes independent of H_2, H_2', and d_2, and its mean value is equal to zero. Then $\mathfrak{M}(d_1 d_2) = 0$ when H_1 is fixed; therefore, for any value of H_1, $\mathfrak{M}(d_1 d_2) = 0$. Thus, $\mathfrak{M}(y_1 y_2) = \Sigma\mathfrak{M}(H_1 H_2) = \phi\Sigma\mathfrak{M}(H^2)$. From this, the correlation coefficient of y_1 and y_2 is

$$r = \frac{\mathfrak{M}(y_1 y_2)}{\mathfrak{M}(y_1^2)} = \frac{\phi}{2(1 + \eta^2 + \xi^2)},$$

where $\eta^2 = \Sigma\mathfrak{M}(d^2)/2\Sigma\mathfrak{M}(H^2)$, and $\xi^2 = \mathfrak{M}(z^2)/2\mathfrak{M}(H^2)$. This can also be written as $r = [(\phi/2)\tau^2]/\sigma^2$, where τ^2 is the "genic additive variance," $2\Sigma\mathfrak{M}(H^2)$, and σ^2 the "total variance," $2\Sigma\mathfrak{M}(H^2) + \Sigma\mathfrak{M}(d^2) + \mathfrak{M}(z^2)$. To avoid having to evaluate ϕ, we note that, since ϕ does not depend on dominance, this formula can be written as $r = r_0\tau^2/\sigma^2$, r_0 being the "fundamental correlation" previously defined. Therefore, for individuals related by one locus only, dominance plays exactly the same role as chance and the environment in reducing all the "fundamental correlations" by a fixed ratio which is less than unity. This formula, in particular, gives, for simple correlations in direct or collateral line of descent, $r = (1/2)^n \tau^2/\sigma^2$, with $n = 1$ for parent-offspring correlation, $n = 2$ for grandparent-grandson, half-sibs, or uncle-nephew, $n = 3$ for first cousins, and so on. It does not apply to full sibs or to double first cousins, who are related by two loci at the same time. We shall show (see §2.5.2) that in these cases, the reduction of the fundamental correlation is less important because there is a positive correlation, $\mathfrak{M}(d_1 d_2) > 0$, between the residual dominance, d_1 and d_2, of the two individuals.

Finally, if we wish to find the partial correlation between trait y in an individual, I_1, and in one of his ancestors, I_2, assuming the value of that trait as fixed in an intermediate ancestor, I_3, who is separated from them by n and p links, respectively, we can apply the following classical formula if the regressions are linear (as they are when the random variables y are Gaussian and in Gaussian relation):

$$
r_{12 \cdot 3} = \frac{r_{12} - r_{13} r_{23}}{\sqrt{(1 - r_{13}^2)} \sqrt{(1 - r_{23}^2)}}
$$

$$
= \frac{(1/2)^{n+p} \tau^2/\sigma^2 - (1/2)^n (\tau^2/\sigma^2)(1/2)^p (\tau^2/\sigma^2)}{\sqrt{(1 - r_{13}^2)(1 - r_{23}^2)}}.
$$

This coefficient, in general positive, is not zero except if $\tau^2/\sigma^2 = 1$, that is, if there is neither dominance nor influence of the environment; it is only in this case that, if we know the trait in an ancestor of I_1, similar information from previous ancestors in the same line of descent would not give us any more information about I_1 (no "ancestral inheritance"). But there is almost always dominance or influence of the environment, and because of this, knowledge about a trait in an ancestor allows a positive correlation among earlier ancestors and the descendants. This "law of ancestral inheritance," shown experimentally by Galton and Pearson, is then not at all in contradiction (as Bateson and Weldon believed) to the laws of Mendel. From Mendel's laws it follows, indeed, that in making predictions about offspring, knowledge of the genetic constitution of one ancestor makes all knowledge about earlier ancestors unimportant. Our study, however, simply shows that knowledge of trait y in a given ancestor, when there is dominance or environmental effects, provides insufficient information about its genetic constitution, and more precise information can be derived from knowledge about earlier ancestors.

2.5.2 Individuals Are Related
by Two of Their Loci

Let $\mathfrak{IC}_1 = H_1 + H_1' + d_1$ and $\mathfrak{IC}_2 = H_2 + H_2' + d_2$. Let us calculate $\mathfrak{M}(\mathfrak{IC}_1\mathfrak{IC}_2)$, knowing that H_1 and H_2 have a correlation coefficient ϕ, that H_1' and H_2' have a coefficient ϕ, and that these two sets of random variables are independent of each other.

The generating function $V(x, y, u, v)$ of all these four random variables taken together is, therefore, the product of the generating functions $V_1(x, y)$ and $V_2(u, v)$ of the two sets H_1 and H_2, H_1' and H_2'.

Let us recall that the generating function of random variables, taking the respective values α, β, and so on, is, by definition, the expectation of $x^\alpha y^\beta \ldots$ (instead of the characteristic function which is the expectation of $e^{\alpha x} e^{\beta y} \ldots$). Therefore,

$$V_1(x, y) = p(p + \phi q)x^t y^t + pq(1 - \phi)(x^t y^s + x^s y^t) + q(q + \phi p)x^s y$$
$$= (px^t + qx^s)(py^t + qy^s) + pq(x^t - x^s)(y^t - y^s),$$

and $V_2(u, v)$ may be expressed in terms of y by replacing x with u, y with v, and ϕ with ϕ'. We know that the generating function of the two variables taken together, $H_1 + H_1'$ and $H_2 + H_2'$, may be obtained by setting $x = u$ and $y = v$ in the product $V_1 V_2$; it is then $W(x, y) = V_1(x, y)V_2(x, y) = \Sigma P_{\alpha\beta} x^\alpha y^\beta$, the coefficient $P_{\alpha\beta}$ of $x^\alpha y^\beta$ in $W(x, y)$ representing, by definition, the probability of also having $H_1 + H_1' = \alpha$ and $H_2 + H_2' = \beta$, and, therefore, of $\mathfrak{IC}_1$ and $\mathfrak{IC}_2$ having determined values $f(\alpha)$ and $f(\beta)$. Knowing W enables us to calculate $\mathfrak{M}(\mathfrak{IC}_1\mathfrak{IC}_2) = \Sigma P_{\alpha\beta} f(\alpha)f(\beta)$ by replacing (in W) x^α by $f(\alpha)$ and y^β by $f(\beta)$, i.e., x^{2t} and y^{2t} by i, and so on. Let us calculate, then,

$$W(x, y) = (px^t + qx^s)^2(py^t + qy^s)^2$$
$$+ pq(\phi + \phi')(px^t + qx^s)(x^t - x^s)(py^t + qy^s)(y^t - y^s)$$
$$+ p^2 q^2 \phi\phi'(x^t - x^s)^2(y^t - y^s)^2;$$

by replacing x^{2t} and y^{2t} by i, x^{t+s} and y^{t+s} by j, x^{2s} and y^{2s} by k, we obtain,

$$\mathfrak{M}(\mathfrak{K}_1\mathfrak{K}_2) = (p^2i + 2pqj + q^2k)^2 + pq(\phi + \phi')(pi + (q - p)j - qk)^2$$
$$+ p^2q^2\phi\phi'(k - 2j + i)^2.$$

This is a symmetric bilinear form of ϕ and ϕ', in which the coefficients are well-determined in a given population and are independent from ϕ and ϕ'. In the same way,

$$r = \mathfrak{M}(y_1y_2)/\sigma y_1\sigma y_2 = \mathfrak{M}(x_1x_2)/\mathfrak{M}(y^2) = \Sigma\mathfrak{M}(\mathfrak{K}_1\mathfrak{K}_2)/\mathfrak{M}(y^2)$$
$$= \{\Sigma pq(\phi + \phi')[pi + (q - p)j - qk]^2 + p^2q^2\phi\phi'(k - 2j + i)^2\}/\sigma^2.$$

Let us calculate the coefficients by giving ϕ and ϕ' specific values. We have seen that, for $\phi' = 0$, r is reduced to $(\phi/2)\tau^2/\sigma^2$. We can write, therefore,

$$r = [(\phi + \phi')\tau^2/2\sigma^2] + \phi\phi'\epsilon^2/\sigma^2,$$

where $\sigma^2 = \mathfrak{M}(y^2) = \Sigma[2\mathfrak{M}(H^2) + \mathfrak{M}(d^2)] + \mathfrak{M}(z^2)$, the total variance; $\tau^2 = 2\Sigma\mathfrak{M}(H^2)$, the genic-additive variance; and $\epsilon^2 = \Sigma\mathfrak{M}(d^2)$, the dominance variance. If we set $\phi = \phi' = 1$, $\mathfrak{K}_1$ and $\mathfrak{K}_2$ become identical, and so we have,

$$\frac{\tau^2 + \epsilon^2}{\sigma^2} = \frac{\Sigma\mathfrak{M}(\mathfrak{K}^2)}{\mathfrak{M}(y^2)} = \frac{\Sigma[2\mathfrak{M}(H^2) + \mathfrak{M}(d^2)]}{\sigma^2}.$$

These calculations, incidentally, lead to,

$$\tau^2 = \Sigma 2pq[pi + (q - p)j - qk]^2$$

and

$$\epsilon^2 = \Sigma p^2q^2(k - 2j + i)^2.$$

Comparing this with the formula $\mathfrak{M}(y_1y_2) = (\phi + \phi')\Sigma\mathfrak{M}(H^2) + \Sigma\mathfrak{M}(d_1d_2)$, we note that $\mathfrak{M}(d_1d_2) = \phi\phi'\epsilon^2 = \phi\phi'\mathfrak{M}(d^2)$; the correlation coefficient between the dominance components d_1 and d_2 of I_1 and I_2 is therefore $\phi\phi'$, which is the product of the correlation coefficients between the genic random variables. It is zero if I_1 and I_2 are related by only one locus, but positive if I_1 and I_2 are related by two of their loci, and this results in an increase of their correlation. For example, for brothers, $\phi = 1/2$, $\phi' = 1/2$, and $r = [(1/2)\tau^2 + (\epsilon^2/2)]/\sigma^2$. This correlation is, therefore, higher than that between parent and off-spring when there is dominance; another reason for this higher correlation is that the effects of environment on two brothers cannot be regarded as independent if they are brought up together. For double cousins, $\phi = \phi' = 1/4$, and $r = [(1/4)\tau^2 + (\epsilon^2/4)]/\sigma^2$; this correlation is higher than that between uncle and nephew.

The phenomenon of dominance is, thus, statistically expressed by correlation coefficients which are higher for the double relationships than for the corresponding simple relationships. This higher correlation decreases rapidly, however, as the relationship becomes more distant, because the product $\phi\phi'$ rapidly becomes negligible.

2.5.3 Various Extensions

(1) The results are valid if there is multiallelism, because $V_1(x, y)$ is still a linear form of ϕ, therefore, $W(x, y)$. $\mathfrak{M}(\mathfrak{IC}_1\mathfrak{IC}_2)$ and r are bilinear symmetrical forms of ϕ and ϕ' whose coefficients are determined by setting $\phi' = 0$, and then $\phi = \phi' = 1$.

(2) The results could be extended to the case where the effects of the different pairs of genes on the traits considered are not additive (generalization of dominance) [3, 11].

(3) The calculations could be modified to take into consideration the resemblance between parents (homogamy); the effect of doing so is to increase all the correlations [3, 11, 22, 23, 24].

(4) If we separate the sexes in the statistical measurements of the correlation, we find, in general, different results for each sex, because of the contribution of sex-linked genes to the trait considered, and the same calculation as in §2.4 could be applied [7, 8].

2.6 CORRELATIONS BETWEEN ANY INDIVIDUALS WITH DOMINANCE

For two individuals, I_1 and I_2, with a coefficient of inbreeding not equal to zero, the calculation of correlations is much less simple when there is dominance, because the four random variables, H_1, H_2,

H'_1, H'_2, will be related among themselves. It then becomes indispensable to determine the generating function of all four random variables, which, for a given type of relationship, can only be done step by step by the following method: given a group of individuals, I_1, I_2, . . . , I_n, let us designate by $P_{\alpha\alpha'\beta\beta'}$ . . . the joint probability that their $2n$ homologous loci are in the states represented by α, α', β, β', . . . (each one of these quantities having one of the values t or s). The generating function for the $2n$ loci will then be:

$$\phi(a_1, a_2, b_1, b_2, . . .) = \Sigma P_{\alpha\alpha'\beta\beta'} . . . a_1^\alpha a_2^{\alpha'} b_1^\beta b_2^{\beta'} . . . ,$$

with the following properties.

If we bring together two groups of individuals with no correlation, the functions ϕ are multiplied.

If we disregard one of the individuals, for example, I_1, the generating function for the remaining individuals can be deduced from ϕ by setting $a_1 = a_2 = 1$.

If we add to the group an offspring, E, from a couple of the group, for example, an offspring of I_1 and I_2, the generating function of the group thus increased will include two more variables, related to E, say, l_1 and l_2; according to Mendel's laws the generating function will then be

$$\Sigma P_{\alpha\alpha'\beta\beta'} . . . \frac{l_1^\alpha + l_1^{\alpha'}}{2} \frac{l_2^\beta + l_2^{\beta'}}{2} a_1^\alpha a_2^{\alpha'} . . . =$$

$$1/4[\phi(a_1 l_1, a_2, b_1 l_2, b_2) + \phi(a_1 l_1, a_2, b_1, b_2 l_2)$$
$$+ \phi(a_1, a_2 l_1, b_1 l_2, b_2) + \phi(a_1, a_2 l_1, b_1, b_2 l_2)].$$

We can then proceed gradually from the probabilities of a given initial group to the probabilities of any group which was derived from it by given matings. The calculations, however, are rarely simple.

Chapter **3**

Evolution of a Mendelian Population

We have discussed, thus far, only a stationary Mendelian population, i.e., a population in which the frequency of any given genes does not change from one generation to the next, a circumstance that can occur only if the population is very large and if the different alleles do not give their carriers either an advantage or a disadvantage in the struggle for existence (i.e., all neutral genes). We shall now consider, first, a population of limited size, and later, a population in which there is selection of genes. We shall see that in such populations the frequency of genes does not remain constant but changes in the course of time. We shall then have to answer two questions: Where does this evolution lead? At what rate does it take place?

3.1 INFLUENCE OF POPULATION SIZE ON NEUTRAL GENES

3.1.1 Constant Population Size

Let us examine a population made up of a constant number of individuals, K, reproducing by random mating, and consider, first, genes whose mutation rate is negligible. Starting with an initial generation, F_0, we designate the successive generations, which we shall assume to be nonoverlapping, by F_1, F_2, and so on; if generations overlap, and mating between different generations is possible, computations become more complicated, but the results are not essentially modified. In spite of random mating, the individuals of the nth generation, F_n, will certainly present some consanguinity if n is sufficiently large, because each one will have at the most K distinct ancestors of order n, rather than the theoretical 2^n ancestors. We could calculate the coefficient of coancestry of an individual only if we knew all the chains of relationship connecting his two parents, i.e., if we had complete records of mating since the beginning. We shall see, however, that one can characterize *a priori* the average coancestry of the nth generation by a number f_n. By definition, f_n will be the probability, evaluated *a priori*, that the two homologous loci of an individual taken at random in F_n are identical, i.e., they come from the same locus of a common ancestor. In each experiment conducted, the *a posteriori* coefficient of coancestry will depend on the individual considered, but given a large number of individuals, f_n will approximate the mean value.

Since the genes being considered are neutral, the *a priori* probabilities of the different alleles will be the same for all generations, that is, p and q, if we assume two alleles only. The formulas $p(p + f_n q)$, $2pq(1 - f_n)$, and $q(q + f_n p)$ will represent the *a priori* probabilities of the three genotypes for the nth generation, and also their frequencies, given enough experiments, in which we would always start with the same frequencies, p and q, for genes A and a. We shall calculate f_n in different cases, disregarding mutations.*

A. Dioecious Individuals. Consider first an animal population with separate sexes, made up of constant numbers N_1 of males and N_2 of females, forming the subpopulations $_1F$ of males and $_2F$ of

---

* The above formulas are based on the assumption that there is only random inbreeding (consistent with panmixia). If there is also systematic inbreeding (see p. 53), the formulas may be modified.

females. Since there is panmixia, the two homologous loci of an individual, I_n, of F_n are taken at random, one from $_1F_{n-1}$, the other from $_2F_{n-1}$. The probability that they come from the same individual of $_1F_{n-2}$ or of $_2F_{n-2}$ is

$$\frac{1}{2}\frac{1}{N_1}\frac{1}{2} + \frac{1}{2}\frac{1}{N_2}\frac{1}{2} = \frac{1}{N}$$

(by designating N as the harmonic mean of $2N_1$ and $2N_2$, $1/N = 1/(4N_1) + 1/(4N_2)$. The complementary probability, $1 - (1/N)$, is the probability that they come from different individuals of F_{n-2}. In these two cases, the probabilities that the two loci are identical are $(1 + f_{n-2})/2$ and f_{n-1} (because f_{n-1} represents the probability that two homologous loci taken from two different individuals of F_{n-2} are identical). Therefore, f_n, the probability that the two loci of I_n are identical, is given by

$$f_n = \frac{1 + f_{n-2}}{2N} + \left(1 - \frac{1}{N}\right)f_{n-1}.$$

From this linear recurrence we can easily deduce f_n. First of all, we return to a homogeneous recurrence by noting that the equation is verified for $f_n = \text{constant} = 1$ and by letting $\alpha_n = 1 - f_n$, from which

$$\alpha_n = [1 - (1/N)]\alpha_{n-1} + (1/2N)\alpha_{n-2}.$$

Here α_n will be a linear combination of two solutions of the form k^n, k being given by the characteristic equation

$$k^2 - [1 - (1/N)]k - 1/2N = 0.$$

Thus,

$$\alpha_n = \lambda[(1 - 1/N + \sqrt{1 + 1/N^2})/2]^n$$
$$+ \mu[(1 - 1/N - \sqrt{1 + 1/N^2})/2]^n,$$

λ and μ being determined by the two initial values of α_0 and α_1,

$$\alpha_0 = \lambda + \mu, \qquad \alpha_1 = \alpha_0[(1 - 1N)/2] + (\lambda - \mu)\sqrt{1 + 1/N^2}/2.$$

Let us note that the indefinite brother-sister mating, studied in detail by Haldane and by Fisher, becomes a special case in this formula if we let $N = 2$:

$$\alpha_n = \lambda[(1 + \sqrt{5})/4]^n + \mu[(1 - \sqrt{5})/4]^n.$$

If N is large, we have

$$\lambda\sqrt{1 + 1/N^2} = [(\sqrt{1 + 1/N^2} - 1 + 1/N)\alpha_0/2] + \alpha_1,$$

$$\lambda = \alpha_1 + \alpha_0/2N.$$

Therefore λ has for its principal part α_1, if $\alpha_1 \neq 0$, that is, if the initial population is not formed by identical homozygotes. The term in μ becomes rapidly negligible with respect to the term in λ, because their ratio is equivalent to $(-1/2N)^n\mu/\lambda$. We have, then, as n starts to increase,

$$\alpha_n = 1 - f_n \sim \alpha_1(1 - 1/2N)^n \sim \alpha_1 e^{-n/2N},$$

and we are led to the following important conclusion: f_n tends toward 1 when n tends toward infinity. Thus, we tend asymptotically toward a population in which the two homologous loci of each individual would have the probability 1 of being identical, and, therefore, a population in which all the loci would be identical, made up of identical homozygotes. For neutral genes and with no mutations, indefinite panmixia in a limited population always leads to complete homogeneity. This result, surprising at first, stems from the fact that a gene can be eliminated when the random drawing of the $2N$ loci of the following generation happens to always favor the same one of the two alleles; on the other hand, a gene, once eliminated, never reappears. The *a priori* probabilities of genes A and a in the F_n generation are certainly always constant and equal to p and q, but this now means that the final population has the probability p of containing only AA's and the probability q of containing only aa's. Here is a large difference from the case of an unlimited population, in which the three genotypes coexist indefinitely with frequencies p^2, $2pq$, q^2. But it must be noted that the asymptotic homogeneity is reached extremely slowly if N is large; for $\alpha_n = 1 - f_n$ to be reduced to one-tenth of its value, a number, n, of generations is required such that $\exp(-n/2N) = 0.1$, and therefore $n = 2N \ln 10$; to appreciably reduce the difference $1 - f$, which measures the deviation from homogeneity, requires about as many generations as there are individuals in the population. These results have important biological consequences; several biologists have insisted on the role of chance in the elimination of neutral genes. We observe, in fact, in many animal and plant species, the divergence of "geographical races," which, after having been separated by a barrier, such as a body of water or a range of mountains, evolve toward different homozygous states, one having finally only genotypes AA, the other having only aa. That divergence could certainly be explained by disruptive selection depending on the geographic situation, the gene A being advantageous in one location, the gene a in another. As frequently happens for neutral genes, however, it must be admitted that this evolution results from a small population becoming homogeneous; this homogenization arises from random elimination, which in its course eliminates sometimes one of the two genes, sometimes the other. This explanation has at times been used improperly; it must be emphasized that random elimination cannot take place in such a short period of time unless the population is very small. Consider the blood group of the American Indians. All these Indians seem to come from the same ancestors, in spite of their morphological variability and limited intermarriage with immigrants from Oceania and Melanesia. They are the only race in the world to have exclusively only one blood group, the group O (OO).

The blood groups A and B, however, result from extremely old mutations, since they probably existed before the separation of the lines of chimpanzees and men, and must have always existed in far eastern Asia, where the great human migrations probably originated. The different blood groups seem to be without selective value, because they coexist in Asia and in Europe under all climates and at all latitudes. It seems, then, that Indians derive from a group of Asiatic immigrants in which the genes *A* and *B* disappeared by random elimination. The group must have developed rapidly, however, after its arrival in the new world; it could not have remained small for more than a few generations, after which the change in gene frequencies had to be very slow. To reach homozygosity within a few generations, the group would have consisted of only a very few individuals. The hypothesis of random elimination of genes *A* and *B* in America thus leads us to consider that most of the American Indians derive genetically from a small number of Asiatics (Mongoloids), who came to America perhaps by crossing over the Bering straits, and to confirm the thesis of American ethnographers, but not the thesis that the American Indian race resulted from hybridization among Mongoloids, Australians, and Melanesians who came at different times by sea. Immigration from Melanesia has had obvious influence only on very isolated regions, such as the Siriono area (the virgin forest of Amazonia). Our hypothesis is further corroborated by the observation that, although the *NN* and *MN* blood groups of the *MM-MN-NN* series occur quite frequently among all races, they hardly ever occur among Indians. Certainly, there are many other genes for which the American Indian population includes heterozygotes, but these genes may have originated, for the most part, from mutations which occurred after the occupation of America. The slowness of random elimination of genes in a population which numbers even a few hundred individuals is confirmed by the example of the Gypsies, nomads who came from India into Europe more than a thousand years ago, and who have conserved remarkably the Hindu type, because they marry almost exclusively among themselves. The few thousand individuals that these isolated populations number in Germany and in France have conserved the same frequency of blood groups as the Hindus, in spite of the thousand-year separation, being 40 per cent *B*, the highest proportion in the world.

B. *Monoecious Individuals.* Consider now a plant population of N monoecious individuals, in which both sexes occur on the same plant. Self-fertilization is now possible, but suppose that it is no more or less probable than cross-fertilization. The two homologous loci of an individual I_n of F_n then have the probability $1/N$ of coming from the same individual of F_{n-1}, in which case their conditional probability of being identical is $(1 + f_{n-1})/2$, and have the probability $1 - 1/N$ of coming from different individuals, in which case their conditional probability of being identical we denote by ϕ_n; then,

$$f_n = [(1 + f_{n-1})/2N] + (1 - 1/N)\phi_n.$$

We have defined ϕ_n as the probability that two homologous loci taken from two different individuals of F_{n-1} are identical. Because of panmixia, $\phi_n = f_{n-1}$; therefore

$$f_n = 1/2N + (1 - 1/2N)f_{n-1},$$

from which

$$\alpha_n = 1 - f_n = (1 - 1/2N)\alpha_{n-1} = \alpha_0(1 - 1/2N)^n.$$

Thus α_n still tends toward zero, and f_n toward 1. The classical case of indefinite self-fertilization is obtained for $N = 1$; $1 - f_n$ then decreases by half in each generation, and almost complete homozygosity is reached quite rapidly. Repeated self-fertilization of a plant species is a rapid procedure for obtaining a line homozygous for almost all factors. But if N is large, homogeneity is established very slowly; α_n is then of the order of $\exp(-n/2N)$, as with dioecious individuals. The occurrence of both sexes on the same plant modifies the evolution of the population to an insignificant extent, provided self-fertilization is not favored more than cross-fertilization, since we have already shown that exclusive self-fertilization rapidly leads to homogeneity.

3.1.2 Population Size Not Constant

Suppose now that population size is not constant but varies over the course of time. Consider the case in which the sexes are separate. The numbers N_1 and N_2 will be functions of the order, i, of the generation F_i. Let us set $1/4N_1 + 1/4N_2 = 1/N(i)$.

The formula

$$f_n = [(1 + f_{n-2})/2N] + (1 - 1/N)f_{n-1}$$

holds, providing we substitute for N the value $N(n - 2)$. From this, by increasing the indices by two to simplify the formula, we deduce

$$\alpha_{n+2} = [1 - 1/N(n)]\alpha_{n+1} + \alpha_n/2N(n).$$

We have this time a linear homogeneous recurrence with variable coefficients. We shall solve it by setting $\alpha_n = k_0 k_1 \ldots k_n$, the k_is being constants to be determined, which are related as follows:

$$k_{n+2}k_{n+1} = [1 - 1/N(n)]k_{n+1} + 1/2N(n);$$

that is,

$$k_{n+2} = [1 - 1/N(n)] + 1/[2N(n)k_{n+1}],$$

which enables us to calculate gradually the k_is, starting with $k_1 > 0$. Then $k_{n+2} > 0$; for $n \geqslant 0$, $k_{n+2} > 1 - 1/N(n)$; and, for $n \geqslant 1$, $k_{n+2} < 1$, because $k_{n+2} < 1$ is equivalent to $k_{n+1} > 1/2$. Therefore the values taken by α_n are positive and decreasing, and, when n tends toward infinity, tend toward a limit, $\alpha \geqslant 0$; and f therefore tends toward the limit $(1 - \alpha) \leqslant 1$.

To make $\lim f = (1 - \alpha) < 1$, that is, for the heterozygotes never to be completely eliminated, it is necessary and sufficient that the series $\log \alpha = \log k_0 + \log k_1 + \ldots + \log k_n + \ldots$ converge. For this k_n must tend toward 1, which necessitates, by the recurrence formula, that both $N(n)$ and n be infinite; this last condition suffices for $k_n \longrightarrow 1$, because, by letting $k_{n+2} = 1 - u_{n+2}$, where $0 < u_{n+2} < 1/N(n)$, the recurrence may be written

$$u_{n+2} = [1/2N(n)][1 - 2u_{n+1})/(1 - u_{n+1})] < 1/2N(n).$$

Therefore $u_{n+2} \longrightarrow 0$ if $N(n) \longrightarrow \infty$. It follows that:

(1) If $N(n)$ remains finite when $n \longrightarrow \infty$, $\lim k_n < 1$, and $\lim \log k_n < 0$; therefore $\log \alpha = -\infty$, $\alpha = 0$, and f tends toward 1.

(2) If $N(n)$ tends toward infinity along with n, $k_n = 1 - u_n \longrightarrow 1$, and $\log k_n = \log(1 - u_n) \longrightarrow 0$. To study the series with general terms $\log k_n$, let us note that

$$u_{n+2} = (1 - \epsilon_n)/2N(n),$$

where

$$\epsilon_n = u_{n+1}/(1 + u_{n+1}) \longrightarrow 0.$$

Therefore

$$u_{n+2} \sim 1/2N(n),$$

and the series $\log k_{n+2} = \log(1 - u_{n+2})$ converges if the series u_{n+2} converges.

(3) If $N(n)$ increases at the most as a linear function of n, the series diverges, and $f \longrightarrow 1$.

(4) If $N(n)$ increases at least as n^{1+k} $(k > 0)$, the series converges, $\lim f < 1$, and there is no complete disappearance of heterozygotes.

The same results would obtain for monoecious plants.

3.1.3 The Role of Mutations

It is obvious that the genetic heterogeneity of a population, i.e., the presence of numerous heterozygotes, does not usually result from the fact that the population is extremely large, but from new genes appearing from time to time, either by mutation or by migration of individuals from a different population. Let u_1 be the mean frequency of mutation per generation for a specified locus, and u_2 the mean frequency of migrants per generation, these migrants coming from a population large enough that we can assume there is no relationship among them. Let us set $u = u_1 + u_2$. The probability that a locus, A_n, comes from a nonmutated locus of an "indigenous" individual (i.e., a nonmigrant) of the preceding generation is $1 - u_1 - u_2 = 1 - u$. As a result, in the case of dioecious individuals,

$$f_n = (1 - u)^4[(1 + f_{n-2})/2N + (1 - 1/N)\phi_{n-1}],$$

ϕ_{n-1} representing the probability that two loci taken from two different indigenous individuals of F_{n-2} are identical. But the coefficient of coancestry, f_{n-1}, of an individual of F_{n-1} is evidently $(1 - u)^2\phi_{n-1}$. From this we deduce

$$f_n = (1 - u)^4(1 + f_{n-2})/2N + (1 - u)^2(1 - 1/N)f_{n-1}.$$

Since we can assume u^2 to be negligible, the equilibrium value of f_n is

$$(1 - 4u)(1 + f)/2N + (1 - 2u)(1 - 1/N)f = \frac{1 - 4u}{4Nu + 1} \sim \frac{1}{1 + 4Nu}.$$

To see how f tends toward this limit, let us set $\alpha_n = f - f_n$. We have

$$\alpha_n = (1 - 4u)\alpha_{n-2}/2N + (1 - 2u)(1 - 1/N)\alpha_{n-1}.$$

This equation has two solutions, $\alpha_n = k_n$, k being a root of

$$k^2 - (1 - 2u)(1 - 1/N)k - (1 - 4u)/2N = 0.$$

Therefore

$$2k = (1 - 2u)(1 - 1/N) \pm \sqrt{(1 - 2u)^2(1 - 1/N)^2 + 2(1 - 4u)/N}.$$

The greater of the two roots is given by:

$$2k = 1 - 2u - 1/N + (1 - 4u - 2/N + 8u/N + 2/N - 8u/N)^{1/2}$$
$$+ 0(u^2) + 0(1/N^2);$$
$$k = 1 - 2u - 1/2N + 0(u^2) + 0(1/N^2).$$

Therefore

$$\alpha_n = f - f_n \sim (1 - 2u - 1/2N)^n \sim e^{-2nu - n/2N} = e^{-(4Nu+1)n/2N}.$$

The equilibrium value is reached more rapidly when there is mutation or migration and much more rapidly if $4Nu$ is large. In monoecious plants, we find again the same results.

In summary, we note that the coefficient of coancestry f_n tends always toward a finite limit f. If it is equal to 1, the population will almost certainly become genetically homozygous, given sufficient time. If it is different from 1, usually some heterozygotes will persist in the final population, the *a priori* probability that an individual taken at random from this population is heterozygous being $2pq(1 - f)$. If the population size cannot be taken as increasing indefinitely, $f = 1/(1 + 4Nu)$ is considerably less than 1, provided that $4Nu$ is not small, or that the frequency u of mutation and migration per generation is on the order of $1/N$, at least, or that the total number, $2Nu$, of new genes introduced in each generation by mutation or migration is one or more. Whatever the population might be, if mutation or migration affect some individuals in each generation, a considerable number of heterozygotes may persist.

3.2 INFLUENCE OF SELECTION

Let us study the distribution, over the course of time, of a pair of factors having only two alleles, A and a.

Let us designate by p and q, $p = 1 - q$, the probabilities of A and a in the adult breeding individuals of the generation F_n. The probabilities $p + \delta p$ and $q + \delta q$ ($\delta q = -\delta p$) in the following generation, F_{n+1}, will be, in general, different from p and q. The change δq results from several causes, each producing a small change (such that their squares and products would be negligible).

A. Mutation. Because of recurrent and reversible mutations, there is, in each generation, in the reproductive cells of F_n a mean proportion, u_1, of a genes transformed to A, and another proportion, v_1, of A genes transformed to a, which gives $q - u_1 q + v_1(1 - q)$ as the average frequency of a in the reproductive cells; the change of q produced by mutation is a linear function of q.

B. Migration. We must take migration into account as soon as we consider a local population instead of all the individuals of a species, since a local population is almost never completely isolated; it always exchanges individuals with the neighboring populations. It follows that, if what we called F_n designates all the indigenous individuals born at a certain place, the breeding population will differ from F_n; it will be formed, on the average, of only a fraction, $1 - k$, of F_n individuals, the remaining fraction, k, consisting of migrant individuals. If we assume that these individuals come from a group of populations whose composition can be considered constant over the course of time and characterized by a frequency, q_m, of the gene a, the mean frequency of a in the breeding population will be

$$(1 - k)q + kq_m = q + k(q_m - q).$$

The change in q caused by migration is, therefore, a linear function of q, as with mutations. It can be written in the same form, $-u_2 q + v_2(1 - q)$, by setting $kq_m = v_2$ and $k = u_2 + v_2$, that is, $u_2 = k(1 - q_m)$. The change produced jointly by mutation and migration is then

$$\delta_1 q = -u_1 q + v_1(1 - q) - u_2 q + v_2(1 - q) = -uq + v(1 - q),$$

by setting

$$u = u_1 + u_2 = u_1 + k(1 - q_m) \quad \text{and} \quad v = v_1 + v_2 = v_1 + kq_m.$$

This change is therefore a linear function of q.

The inclusion of both mutation and migration effects in the same formula is based on a simplified, gross model which assumes that the migrants come from outside populations whose composition remains constant in time. In reality, these populations undergo evolution, and are themselves affected by migration. What should be studied, then, is the evolution of a group of populations interacting with each other by migration (see §3.3).

For the time being we shall assume that the reproductive cells of the F_n generation, because of migration, carry the gene a with the frequency $q_1 = q + \delta_1 q = q - uq + v(1 - q)$.

C. Gametic Selection. Assume that the gametes produced by these reproductive cells do not carry the gene a with a probability q_1 any longer, but with a different probability, q_2, because this gene presents an advantage or a disadvantage for the gametes which carry it (gametic selection); assume, also, that the probabilities of the two genes, instead of being q_1 and $1 - q_1$, are $q_2 = \alpha q_1$ and $1 - q_2 = \beta(1 - q_1)$, α and β having a constant ratio close to 1, designated by $1 - s$. This ratio characterizes the degree of viability of the gametes, i.e., the intensity of "gametic selection" (we can, if necessary, consider s positive, by calling a the unfavorable gene).

Since we must have

$$\alpha q_1 + \beta(1 - q_1) = 1, \quad \text{with} \quad \alpha/\beta = 1 - s,$$

we have

$$\beta(1 - s)q_1 + \beta(1 - q_1) = 1,$$
$$1/\beta = 1 - sq_1,$$

and

$$\alpha \sim 1 - s + sq_1 + 0(s^2).$$

Let us put $q_2 = \alpha q_1 = q_1 + \delta_2 q$. Then $\delta_2 q = (\alpha - 1)q_1 = -sq_1(1 - q_2)$.

D. Consanguinity. Let us assume pure consanguinity, because of which each gamete contributing to reproduction, whatever its constitution may be, has the same probability of uniting with another gamete; the eventual consanguinity, however, will increase the probability that the other gamete carries the same gene as the first one. If we call f the average coefficient of inbreeding of generation F_{n+1}, the gametes that unite to form the individuals born, or "zygotes" of generation F_{n+1}, have among themselves on the average a conditional* correlation coefficient l, and each of them carries A or a with the probabilities p_2 and q_2; the three zygotes AA, Aa, aa, have then the probabilities

$$P = p_2(p_2 + lq_2), \qquad 2Q = 2p_2q_2(1 - l), \qquad R = q_2(q_2 + lp_2).$$

E. *Zygotic Selection.*   Assume that the three zygotes do not have the same probability of developing and reaching the adult reproductive stage and that the probabilities of the adults are νP, $2\mu Q$, and γR. The three quantities ν, μ, and γ have constant ratios close to 1: $\frac{\gamma}{\nu} = 1 - \sigma; \frac{\mu}{\nu} = 1 - h\sigma$. The two constants σ and h characterize the degree of viability of the zygotes, or the intensity of "zygotic selection." The heterozygotes will be intermediate in viability between the two homozygotes if $0 < h < 1$; they will be superior to both homozygotes if $h < 0$ (and inferior to both if $h > 1$) and if $\sigma > 0$, and vice versa if $\sigma < 0$. We must have

$$\nu P + 2\mu Q + \gamma R = 1,$$
$$\nu[(1 - \sigma)R + 2(1 - h\sigma)Q + P] = 1,$$

from which

$$\nu = 1/(1 - \sigma R - 2h\sigma Q) = 1 + \sigma R + 2h\sigma Q + O(\sigma^2).$$

The probability $q_2 + \delta_3 q$ of a in the adult breeding individuals of F_{n+1} is therefore

$$q_2 + \delta_3 q = \mu Q + \gamma R$$
$$= \nu[(1 - h\sigma)Q + (1 - \sigma)R]$$
$$= (q_2 - h\sigma Q - \sigma R)/(1 - \sigma R - 2h\sigma Q),$$

and by grouping the terms in Q and R and reducing the denominator to 1,

---

* This *conditional* correlation coefficient, which equals zero in case of random mating, is not the same as the a priori coefficient of §3.1, which was larger than zero.

$$\delta_3 q = h\sigma p_2 q_2 (1 - l)(2q_2 - 1) - \sigma p_2 q_2 (q_2 + l p_2) + O(\sigma^2)$$

$$= \sigma p_2 q_2 [(2h - 1)(1 - l)q_2 + hl - l - h] + O(\sigma^2),$$

we can replace q_1 by q_2 and p_2 by p_1, which differ by just $O(s\sigma)$ [since $q_2 = q_1 + O(s)$]. We note then that the total change, $\delta_2 q$ and $\delta_3 q$ due to gametic and zygotic selection, is of the form

$$\delta_2 q + \delta_3 q = q_1 (1 - q_1)(t + wq_1)$$

(disregarding the second-order terms in s and σ), with

$$t = -s - f\sigma - h\sigma(1 - l), \quad \text{and} \quad w = \sigma(2h - 1)(1 - l).$$

If s, σ and $h\sigma$ have the same sign, t will also have the same sign; t will be called the coefficient of total selection. These two selections together produce, therefore, a change which is a third-degree function in q, becoming equal to zero for $q = 0$ and $q = 1$. In fact, the selection ceases to operate when a is eliminated ($q = 0$) or is fixed ($q = 1$). An essential difference exists between the above change and the change produced by mutations or migration. The latter is a first-degree function, and does not become equal to zero at the limits because it is always affected by a gene even if it is fixed or eliminated.

The function is reduced to the second degree if $w = 0$, i.e., if $\sigma = 0$ (selection exclusively gametic), or if $h = 1/2$ (heterozygote exactly intermediate from the point of view of viability), or if $\lambda = 1$ (population consisting of homozygotes, exclusively).

We can replace q_1 by q in the formula if the second-order terms in u, v, s, and σ are disregarded. We then obtain for the total change, $\delta q = \delta_1 q + \delta_2 q + \delta_3 q$, in one generation,

$$\delta q = \underbrace{-uq + v(1 - q)}_{\substack{\text{mutation} \\ \text{and migration}}} + \underbrace{q(1 - q)(t + wq)}_{\text{selection}}.$$

This third-degree polynomial we will call $\delta(q)$. Its coefficients are so small that their products and squares can be disregarded.* It gives the changes in the probability q of a, as well as in its frequency in a very large population, caused by mutation, migration, and selection.

3.2.1 The Case of a Very Large Population

The difference between the probability and the frequency of a gene in a very large population is negligible. The frequency q varies from one generation to the next by a quantity $\delta(q)$, supposed to be small; q is a function of time, measured in generations, whose finite difference is the function $\delta(q)$. The integration of $\delta(q)$ is approximately reduced to the following quadratic form:

$$\frac{dq}{dt} = \delta(q), \qquad t = \int_{q_0}^{q} \frac{dq}{\delta(q)}$$

q_0 being the initial value in the generation $t = 0$.

We shall proceed to obtain the limit of q when $t \longrightarrow +\infty$ by a graphic discussion (asymptotic distribution of the genes). Let us assume that the rates of mutation and migration, u and v, are not equal to zero. We note then that the third-degree polynomial $\delta(q)$ is equal to $v > 0$ for $q = 0$ and to $-u < 0$ for $q = 1$. It does not reduce to zero for $q = 0$ or for $q = 1$, and it allows either one or three intersections between 0 and 1. To be more specific,

$$\frac{\delta(q)}{q(1-q)} = t + wq - \left(\frac{u}{1-q} - \frac{v}{q}\right) = y_1 - y_2.$$

Let us represent in a plane $(\bar{q}, y)$ the straight line D, generated by $y_1 = t + wq$, and the curve C, generated by

$$y_2 = \frac{u}{1-q} - \frac{v}{q}, \qquad 0 < q < 1.$$

---

* We could, of course, formulate $\delta(q)$ without making these approximations, but the expression obtained would be unmanageable except in such special cases as the study of lethal genes by Teissier [21], where aa is nonviable and $\sigma = 1$; σ^2 would not be negligible, but the formula would nevertheless be simplified, because there would be only two genotypes present.

The curve rises from the point $(0, -\infty)$ to the point $(1, +\infty)$, because

$$y_2' = \frac{u}{(1-q)^2} + \frac{v}{q^2}.$$

It crosses the x-axis at the point $q_1 = \dfrac{v}{u+v}$. (See Figure 5.)

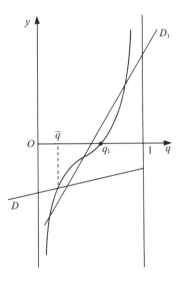

FIGURE 5.

(A) Let us assume that curve C meets the straight line D in only one point, Q, of the abscissa, $\bar{q}$. Since $\delta(\bar{q}) = 0$, an initial frequency, q_0, that was equal to $\bar{q}$ would remain constant through the generations (stationary frequency). In the general case $y_1 - y_2$, therefore, $\delta(q) > 0$ if $q < \bar{q}$ and < 0 if $q > \bar{q}$; $\delta(q)$ is, therefore, always opposite in sign to $q - \bar{q}$. The difference $q - \bar{q} = r$ decreases constantly in absolute value from its initial value, $r_0 = q_0 - \bar{q}$; to see if it tends toward zero, and at what rate, let us study the quotient $\dfrac{-\delta(q)}{q - \bar{q}}$, which is a polynomial of at most the second degree, positive, and never equal to zero. Let us call $m > 0$ its minimum in the range of values taken by q, i.e., between q_0 and $\bar{q}$; if q_0, and consequently q, is sufficiently close to $\bar{q}$, one could write essentially $\delta(q) = \delta'(\bar{q})(q - \bar{q})$, and, thus, take approximately $m = \delta'(\bar{q})$. Thus, with $\Delta r = r' - r$ designating the change in r from one generation to the next, we have

$$\frac{-\Delta r}{r} > m, \qquad \frac{-\Delta|r|}{|r|} > m, \qquad \Delta|r| < -m|r|,$$

$$|r'| = |r| + \Delta|r| < (1 - m)|r|.$$

After n generations, $|r| < (1 - m)^n|r_0|$; therefore, $r = q - \bar{q}$ tends toward zero at least as fast as $(1 - m)^n$ does. The stationary frequency $q = \bar{q}$, considered earlier, is stable, and any other frequency tends asymptotically toward it, the deviation $r = q - \bar{q}$ being multiplied after n generations by a quantity certainly less than $(1 - m)^n$.

There are two important specific cases.

(1) In the first one there is no selection; mutations and migration act alone; $w = t = 0$; and D coincides with the x-axis. The asymptotic value, $\bar{q}$, is equal to

$$q_1 = \frac{v}{u + v} \qquad (q_1 = 1/2 \text{ if } u = v);$$

$$\delta(q) = -uq + v(1 - q) = -(u + v)(q - \bar{q}).$$

Therefore, $m = u + v$; $q - \bar{q}$ is reduced in n generations to a quantity less than $(1 - u - v)^n$. This reduction is not significant unless n is on the order of $\dfrac{1}{u + v}$; if u and v are reduced to the rate of mutation, which is extremely low (on the order of 10^{-5}), $\bar{q}$ does not noticeably approach the asymptotic value unless the number n of generations is on the order of 10^5. It will be almost impossible to observe a population that became stationary under the action of mutations alone. Moreover, the irregularity in the rate of mutations, as well as in the rate of migration, restricts the validity of the formula, but in practice selection usually plays the principal role.

(2) In the second case there is gametic selection only, with heterozygotes being exactly intermediate in viability; $w = 0$; D is hori-

zontal, with ordinate t (coefficient of total selection); $t < 0$ if the gene a is selected against; and q tends toward the asymptotic value $\bar{q}$, which is lower than $q_1 = \dfrac{v}{u + v}$. Let us calculate $\bar{q}$. We have

$$\delta(q) = -tq^2 + (t - u - v)q + v,$$

the roots of which are

$$\frac{-t + u + v \pm \sqrt{(t - u - v)^2 + 4vt}}{-2t}.$$

Since $\delta(1) < 0$, therefore $\bar{q}$, which lies between 0 and 1, is the smallest root; the other root, $\bar{\bar{q}}$ is obtained by taking the positive value of the radical, and we have

$$\delta(q) = -t(q - \bar{q})(q - \bar{\bar{q}}).$$

Therefore we will take for m, the minimum of $\dfrac{-\delta(q)}{q - \bar{q}}$, the minimum of $-t(\bar{\bar{q}} - q)$, which is the smallest of the two quantities $-t(\bar{\bar{q}} - \bar{q})$ and $-t(\bar{\bar{q}} - q_0)$.

In the specific and usual case where u and v (reduced to the mutation rate without any migration) are small compared with the coefficient of total selection, t, the roots are given by

$$\frac{-t + u + v}{-2t}\left[1 \pm \sqrt{1 + \frac{4vt}{(-t + u + v)^2}}\right],$$

which is equivalent to $(1/2)\left[1 \pm \left(1 + \dfrac{2v}{t}\right)\right]$; therefore $\bar{q} \sim -v/t$, $\bar{\bar{q}} \sim 1 + v/t$, and the asymptotic value $\bar{q} = -v/t$ is small. Selection eliminates almost completely the unfavorable gene a; its complete disappearance is prevented by the mutation rate, v, alone. Unless q_0 is not close to $\bar{\bar{q}}$, i.e., close to 1, m is on the order of $-t$, and would not equal $u + v$ unless there was selection; the asymptotic value is, therefore, reached much more rapidly.

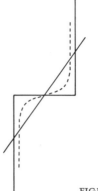

FIGURE 6.

(B) The curve C can be met by a straight line, such as D_1, in three points (see Figure 6), when C has two real tangents with the same slope, w, as D_1, and when D_1 falls between these two tangents. The tangents parallel to D_1, then, have their points of contact given by $\dfrac{u}{(1-q)^2} + \dfrac{v}{q^2} = w$, an equation having two real solutions, q, between 0 and 1 if w is greater than the minimum, $(u^{1/3} + v^{1/3})^3$, of the first member between 0 and 1. If, moreover, t is within the interval $t_1 \ldots t_2$ from the ordinates to the origin of the tangents, the equation $\delta(q) = 0$ will have three solutions between 0 and 1, in order of magnitude $\bar{q}_1$, $\bar{q}_2$, $\bar{q}_3$. Each of these values results in a stationary distribution that is maintained indefinitely, but if we start with a different value of q_0, Figure 6 shows that:

(1) If $q_0 < \bar{q}_2$, $\delta(q) = y_1 - y_2$ is opposite in sign to $q - \bar{q}_1$; the difference $r = q - \bar{q}_1$ decreases in absolute value from its initial one, $r_0 = q_0 - \bar{q}_1$; if we take $m > 0$ as the minimum of $\dfrac{-\delta(q)}{q - \bar{q}}$ in the interval $q_0 \ldots \bar{q}_1$, the difference r is still reduced after n generations by a quantity less than $(1 - m)^n$, and q tends toward the asymptotic value $\bar{q}_1$.

(2) If $q_0 > \bar{q}_2$, the same reasoning shows that q tends toward the asymptotic value $\bar{q}_3$. The intermediate root, $\bar{q}_2$, of $\delta(q)$ corresponds,

therefore, to an unstable stationary state, which, depending on whether q_0 is smaller or larger than $\bar{q}_2$, tends toward the stable stationary values $\bar{q}_1$ or $\bar{q}_3$.

(C) Let us study directly any type of selection, when mutations and migration are negligible, i.e., $u = v = 0$; this case does not come directly under the preceding presentation, because under these conditions, curve C degenerates. We have

$$\delta(q) = q(1 - q)(t + wq) = wq(1 - q)(q - \alpha), \quad \text{with } \alpha = -t/w$$

(α could be inside or outside the interval $0 \dots 1$). The stationary values are $q = 0$, $q = 1$, and $q = \alpha$ if $0 < \alpha < 1$.

(1) If $\alpha > 1$ or $\alpha < 0$, $\delta(q)$ has a constant sign; if, for example, the sign is negative, q always decreases; $-\delta(q)/q$ has a positive minimum m. We deduce from this that q tends toward zero faster than $(1 - m)^n$ does, and so gene a is eliminated whatever its initial frequency was (if there were a very low rate of mutation, a would persist with a low frequency, as happens with gametic selection). If $\delta(q)$ is positive, $q \longrightarrow 1$ and gene a is fixed whatever its initial frequency was.

(2) If $0 < \alpha < 1$, two cases must be distinguished:

(a) If $w > 0$, $\delta(q)$ always has the same sign as $q - \alpha$. The change in q, and therefore in $q - \alpha$, has the same sign as $q - \alpha$; $q - \alpha$ increases in absolute value from its initial value of $q_0 - \alpha$. As previously, we note that q tends toward zero if $q_0 < \alpha$, and q tends toward 1 if $q_0 > \alpha$. One of the genes is still eliminated, but this time which gene is eliminated depends on the initial frequency.

(b) If $w < 0$, $\delta(q)$ is always opposite in sign to $q - \alpha$. We note again that the difference $r = q - \alpha$ decreases in absolute value and tends to zero. In the asymptotic distribution, the two genes a and A coexist with the stationary frequencies α and $1 - \alpha$. It is easy to see that this is so in the important, specific situation when $s = 0$, $\sigma < 0$, $h > 1$, there is exclusively zygotic selection, and the heterozygote is superior in viability to either homozygote, provided consanguinity is not too high. In fact, we have $w < 0$, and $\alpha = -t/w =$

$\left(h + \dfrac{\lambda}{1 - \lambda}\right) \Big/ (2h - 1)$ is positive but not less than 1 except if

$\dfrac{\lambda}{1 - \lambda} < (h - 1)$, which makes necessary that $1/(1 - \lambda) < h$, that

is, $\lambda < 1 - 1/h$.

REMARK.
We can easily verify that the case $u = v = 0$ of (C) appears as a special case in the graphic discussion of (A) or of (B), if we consider the curve C to have degenerated into the broken line defined by ($q = 0$, $y < 0$; $0 < q < 1$, $y = 0$; $q = 1$, $y > 0$).*

It follows that if u and v are small with respect to t and w but not equal to zero (dotted line), the discussion will be the same as in (C), the only difference being that elimination and fixation will be replaced by an asymptotic equilibrium corresponding to a frequency of $\bar{q}$, close to 0 or 1.

3.2.2 The Case of a Finite Population

Let N be the number of individuals in each generation. If q is the frequency of a in F_n, we have seen that the probability of a in F_{n+1} will be $q + \delta(q)$, $\delta(q)$ being represented (as a first approximation) by a third-degree polynomial. But the frequency, q_1, of a in F_{n+1} will differ from the probability, $q + \delta(q)$, because this frequency is a random variable for which $q + \delta(q)$ represents only the mean value. When the law of probability of q_1 is known as a function of q, e.g., $\theta(q, q_1) \, dq_1$, the frequencies of a in successive generations appear as random variables in the simple Markov chain whose law of transition is $\theta(q, q_1) \, dq_1$, which is assumed to be independent of the rank, n, of the generation considered, as is possible if N is constant. If we assume that the transition in one generation, or in a certain number of generations, of any frequency, q, to any other frequency,

---

* C will be met at only one point by the straight line D if $\alpha > 1$ or $\alpha < 0$ (case 1) or if $0 < \alpha < 1$ and $w < 0$ (case 2b), but in three points if $0 < \alpha < 1$ and $w > 0$ (case 2a).

q_1, is possible, then $\theta(q, q_1)$ is always greater than zero. This assumption implies that the rates of mutation u and v are not equal to zero, because otherwise we could not pass from $q = 0$ or $q = 1$ to different values. Markov's theorem indicates then that the *a priori* law of probability, $\phi_n(q)\, dq$, of the frequency of q in the generation F_n tends toward a limit law, $\phi(q)\, dq$, which is independent of the initial value of q, when n tends toward infinity.

It is possible to formulate these laws in terms of certain hypotheses concerning the law of transition, $\theta(q, q_1)\, dq_1$, which is the law of probability of q_1 when q is fixed. Let us assume it to be a form of Gauss's law with mean value $q + \delta(q)$, $\delta(q)$ being small and such that $\delta(0) \geqslant 0$ and $\delta(1) \leqslant 0$, and with a small variance, $\sigma^2 = w(q) \geqslant 0$, being equal to zero only for $q = 0$ and $q = 1$.

Let us assume, for instance, that the $2N$ gametes which produce the F_{n+1} are taken at random from an infinitely large number of gametes produced by F_n and have essentially the frequencies q and $(1 - q)$ for a and A. We know that the law of probability of the frequency of a in F_{n+1} will be practically Gaussian, and that the conditional variance of this frequency with respect to its mean value will be $\sigma^2 = w(q) = q(1 - q)/2N$, which does not quite become zero except if $q = 0$ or $q = 1$.

In general, if because of systematic consanguinity in F_n the N zygotes of F_{n+1} are each taken at random with the conditional probabilities $P = p(p + \lambda q)$, $2Q = 2pq(1 - \lambda)$, and $R = q(q + \lambda p)$ for the three states AA, Aa, and aa, λ being the conditional inbreeding coefficient of F_{n+1}, the conditional variance of the frequency q_1 of a in F_{n+1} with respect to its mathematical expectation q will be $\sigma^2 = q(1 - q)(1 + \lambda)/2N$.

A. Fundamental Equation. In the transition from generation F_n to F_{n+1}, the *a priori* law of probability of the frequency changes from $\phi_n(q)\, dq$ to

$$\phi_{n+1}(q_1)\, dq_1 = dq_1 \int_0^1 \phi_n(q)\theta(q, q_1)\, dq.$$

If we call M_i and M_i' the moments of the *a priori* law of probability in F_n and in F_{n+1}, we have:

$$M_i = \int_0^1 q^i \phi_n(q)\, dq;$$

$$M_i' = \int_0^1 q_1^i \phi_{n+1}(q_1)\, dq_1$$

$$= \int_0^1 \left[\int_0^1 q_1^i \theta(q, q_1)\, dq_1 \right] \phi_n(q)\, dq$$

$$= \int_0^1 \mu_i(q)\phi_n(q)\alpha q$$

(by inverting the integrations, which is legitimate for functions that are bounded and can be integrated within finite intervals).

If $\mu_i(q)$ are the moments of Gauss's law, $\theta(q, q_1)\, dq_1$, whose mean and variance are $q + \delta(q)$ and $w(q)$, respectively, and if δ and w are small, these moments are calculated by developing the characteristic function according to the powers of its variable τ:

$$\exp\left[(q + \delta)\tau + w\tau^2/2\right] = 1 + (q + \delta)\tau + w\tau^2/2! + \ldots$$
$$+ \left[(q + \delta)\tau + w\tau^2/2\right]^i/i! + \ldots.$$

We see that, by disregarding the terms in w^2 and δ^2,

$$\mu_i/i! = (q + \delta)^i/i! + (i - 1)(q + \delta)^{i-2}w/2(i - 1)! + 0(w^2),$$

and

$$\mu_i = q^i + i\delta q^{i-1} + \frac{i(i - 1)}{2} wq^{i-2} + 0(w^2) + 0(\delta^2) + 0(w\delta);$$

therefore, the variance of the moments from one generation to the next is

$$M_i' - M_i = \int_0^1 [\mu_i(q) - q^i]\phi_n(q)\, dq$$

$$= i\int_0^1 \delta(q)q^{i-1}\phi_n(q)\, dq + \frac{i(i - 1)}{2}\int_0^1 w(q)q^{i-2}\phi_n(q)\, dq.$$

If we assume that $\delta(q)$ and $w(q)$ can be represented by polynomials, verified exactly by the specific forms which we have indicated, we shall write:

$$\delta(q) = \sum_{l \geqslant 0} Alq^l; \qquad w(q) = \sum_{l \geqslant 0} Blq^l.$$

By comparing the small variance, $M_i' - M_i$, to a derivative dM_i/dt (time, t, being measured in generations), equation (3.2.1) is transformed to a differential system for the moments:

$$\frac{dM_i}{dt} = i\Sigma A_l M_{i-1+l} + \frac{i(i-1)}{2} \Sigma B_i M_{i-2+l}. \qquad (3.2\ 2)$$

This system cannot be solved directly, because in the second part of the equation there are moments of higher order than in the first; it enables us, however, to obtain a partial derivative equation for the characteristic function (or Laplace transformation) of the probability law $\phi(q, t) \, dq$,* for which the moments are $M_i(t)$. In fact, this transformation is

$$F(s, t) = \int_0^1 e^{sq}\phi(q, t) \, dq = \sum_{p \geqslant 0} M_p(t)s^p/p!,$$

with derivative

$$\frac{\partial^k F}{\partial s^k} = \Sigma M_p s^{p-k}/(p - k)!;$$

these functions always exist since we integrate only between 0 and 1. By multiplying equation (3.2.2) by $s^{i-1}/i!$ and summing over i from 0 to $+\infty$, we obtain

$$\frac{1}{s}\frac{\partial F}{\partial t} = \Sigma A_l \frac{\partial^l F}{\partial s^l} + \frac{s}{2} \Sigma B \frac{\partial^l F}{\partial s^l}. \qquad (3.2.3)$$

Following the Laplace transformation, by setting

$$F(s, t) = \mathcal{L}[\phi(q, t)],$$

---

* This function should be integrable when $0 \leqslant q \leqslant 1$ for all values of t. It will be evident from equations (3.2.6) and (3.2.10) that the condition has to be supposed true only for $t = 0$, provided $u > 0$ and $v > 0$.

we have

$$\frac{\partial F}{\partial t} = \mathcal{L}\left[\frac{\partial \phi}{\partial t}\right],$$

$$\frac{\partial^l F}{\partial s^l} = \int_0^1 e^{sq} q^l \phi(q, t) \, dq = \mathcal{L}[q^l \phi(q, t)],$$

$$\frac{1}{s}\frac{\partial F}{\partial t} = \frac{1}{s}\int_0^1 e^{sq}\frac{\partial \phi}{\partial t}\, dq = \frac{1}{s}\int_0^1 e^{sq}\frac{\partial V}{\partial q}\, dq$$

$$= -\int_0^1 e^{sq} V \, dq = -\mathcal{L}[V],$$

by setting

$$V = \int_0^q 0\frac{\partial \phi}{\partial t}\, dq = \frac{\partial}{\partial t}\left[\int_0^q \phi \, dq\right],$$

and noting that $V = 0$ for $q = 0$ and for $q = 1$.

Since two functions for which the Laplace transformations are the same are identical almost everywhere, we obtain from equation (3.2.3):

$$-\frac{\partial}{\partial t}\left[\int_0^q \phi(q, t)\, dq\right] = \Sigma A_l q^l \phi(q, t) - 1/2\frac{\partial}{\partial q}[\Sigma B_l q^l \phi(q, t)];$$

that is,

$$\frac{\partial}{\partial t}\left[\int_0^q \phi(q, t)\, dq\right] = (1/2)\frac{\partial}{\partial q}[w(q)\phi(q, t)] - \delta(q)\phi(q, t). \qquad (3.2.4)$$

Such is the fundamental equation.

B. *Asymptotic Probability Law.* If we consider $\phi(q) \, dq$ the law of asymptotic probability for infinite t, then, according to Markov's theory, the law of stationary probability, verifying (3.2.4), will be

$$(1/2)\frac{\partial}{\partial q}[w\phi] - \delta\phi = 0. \qquad (3.2.5)$$

It is, therefore, the law for which the probability density is

$$\phi(q) = [K/w(q)]e^{2 \int \frac{\delta(q)}{w(q)} dq}. \tag{3.2.5'}$$

In particular, when

$$w = q(1 - q)/2N$$

and

$$\delta(q)/q(1 - q) = -u/(1 - q) + (v/q) + t + wq,$$

we have

$$\phi(q) = K_1 q^{4Nv-1}(1 - q)^{4Nu-1} e^{2N(wq^2+2tq)dq}, \tag{3.2.6}$$

with K_1 determined in such a way that the integral between 0 and 1 is equal to 1.

This formula, given by Wright [22, 23, 24] for specific cases but without general demonstration, represents the probability that, in a limited population of N individuals, a gene a, with given coefficients of mutation, migration, and selection, after an infinitely large number of generations, has a frequency between q and $q + dq$. It also represents, therefore, the law of asymptotic distribution of gene a, after an infinitely long time in an infinitely large number of populations of the same size N, and in which all the coefficients would be the same. Let us indicate some specific cases.

(1) If $u = 0$, or $v = 0$, K_1 is by necessity zero, since the integral between 0 and 1 of $1/q$ or of $1/(1 - q)$ is infinite. This result indicates that, eventually, genes not affected by mutation or migration will certainly be either eliminated or fixed.

(2) If $4Nu$ and $4Nv$ are less than 1, i.e., if the population size is large enough, and the mutation or migration rates are not too low, $\phi(q) = 0$ for $q = 0$ and $q = 1$, and is represented by a bell- or double-bell-shaped curve (Figure 7) with one or more dominant q_1 given by the equation $\frac{\partial \phi}{\partial q} = 0$, that is, $4N\delta(q_1) + 2q_1 - 1 = 0$, which, for a very large N, becomes $\delta(q) = 0$; i.e., it gives again the same

58

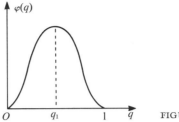

FIGURE 7.

equation or the asymptotic values $\bar{q}$ in a very large population that were studied in §3.2.1.

Simple results are obtained by assuming that there is only one dominant q_1, and that q remains close to it with a probability not far from 1; as a first approximation, let us replace $\delta(q)$ by a linear function of q, i.e.,

$$\delta(q) = -k(q - \bar{q}),$$

$\bar{q}$ being, by necessity, the asymptotic value in a very large population, as in §3.2.1(A), 3.2.1(C.1), or 3.2.1(C.2.b), and k being equal to $-\delta'(\bar{q})$ and therefore being on the order of magnitude of the largest of the numbers u, v, w, t, according to Taylor's formula. Formula (3.2.7) would be exact if $\delta(q)$ was linear, i.e., if there was no selection, as in §3.2.1(A.1). The dominant q_1 is given, then, by

$$-4Nk(q_1 - \bar{q}) + 2q_1 - 1 = 0,$$

from which

$$q_1 = \frac{4Nk\bar{q} - 1}{4Nk - 2} \sim \bar{q},$$

and if $4Nk$ is large, the dominant coincides essentially with the asymptotic value in an infinite population. The asymptotic distribution (3.2.5') is written

$$\phi(q) = K_1 q^{4Nk\bar{q} - 1}(1 - q)^{4Nk(1 - \bar{q}) - 1}$$

with

$$K_1 = B(4k\bar{p}, 4k\bar{q}),*$$

and its moments are given by formula (3.2.2), which can be written

$$0 = i(-kM_i + k\bar{q}M_{i-1}) + \frac{i(i-1)}{4N}(M_{i-1} - M_i).$$

If we start with $M_0 = 1$, $M_1 = \bar{q}$, and

$$2k(M_2 - \bar{q}^2) = \frac{1}{2N}(\bar{q} - M_2),$$

from which we can deduce (σ^2 being the variance) that

$$\sigma^2 = M_2 - \bar{q}^2 = \frac{\bar{q}(1 - \bar{q})}{4Nk + 1} \sim \frac{\bar{q}(1 - \bar{q})}{4Nk}$$

if $4Nk$ is large, i.e., if the order of magnitude of k is greater than that of $1/N$; σ^2 is then small, the distribution is concentrated, and it is legitimate to admit that q varies, practically, in an interval of limited range and that $\delta(q)$ is linear.

For a large number of populations under the same conditions, all having the same size, N, sufficiently large for $4Nk$ to be large, the asymptotic frequencies observed will almost all be grouped around the value that corresponds to an infinite population. The experimental estimate of the variance of these frequencies will enable us to determine k, if we know N, and the variance of $\bar{q}$ gives $v/(u + v)$ or $-(v/t)$ or $[h + (\lambda/1) - \lambda]/(2h - 1)$.

(3) If $4Nu$ and $4Nv$ are less than one, $\phi(q)$ is infinite for $q = 0$ and $q = 1$ and is represented by a U-shaped curve (see Figure 8). The smaller u and v get, the smaller K_1 becomes. It is the frequencies close to $q = 0$ and $q = 1$ that have by far the highest probability. Most of the genes are approaching fixation or elimination, and the only thing that stops the approach is recurring mutations or renewed migration. There is a basic difference, then, between the case of a population that is very small or that has very low rates of mutation

---

* B stands for the Eulerian integral.

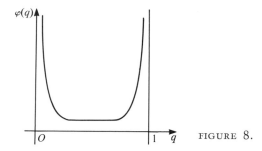

FIGURE 8.

and migration, and whose genes tend toward fixation or elimination, and the case of a large population with each gene almost stabilized around a determined frequency.

C. Evolution of the Probability Law over Time.

In verification of equation (3.2.4), let us call $\phi(q, t)$ and $\Phi(q, t) = \int_0^q \phi(q, t) \, dq$ the law of elementary probability and the integral at time t, respectively; let us call $\phi(q)$ and $\Phi(q) = \int_0^q \phi(q) \, dq$ the asymptotic law, deduced from (3.2.5); we designate by $R(q, t) = \Phi(q, t) - \Phi(q)$ the difference between the integral law and the asymptotic law at instant t. This difference is given for the initial instant as $R(q, 0) \equiv R_0(q)$; it satisfies conditions at the limits $R(0, t) \equiv R(1, t) \equiv 0$ and it verifies, evidently, the equation [obtained while deriving (3.2.5) from (3.2.4)]

$$\frac{\partial R}{\partial t} = \frac{1}{2} \frac{\partial}{\partial q} \left[w(q) \frac{\partial R}{\partial q} \right] - \delta(q) \frac{\partial R}{\partial q}. \tag{3.2.7}$$

The difference will be determined, therefore, by obtaining the solutions of (3.2.7) which become zero for $q = 0$ and $q = 1$ and are of the form $R \equiv K(q) \cdot L(t)$. These solutions must satisfy

$$\frac{L'(t)}{L(t)} = \frac{w}{2} \frac{K''(q)}{K(q)} + \left(\frac{w'}{2} - \delta \right) \frac{K'(q)}{K(q)},$$

for which it is necessary that

$$L(t) = e^{-\lambda t}, \tag{3.2.8}$$

and

$$\frac{wK''}{2} + \left(\frac{w'}{2} - \delta\right) K' + \lambda K = 0, \qquad (3.2.9)$$

with $K(0) = K(1) = 0$.

The last equation cannot be satisfied with these conditions at the limits unless the constant λ belongs to the series of "proper values," λ_i, which are real, positive, and, presumably, arranged in increasing order of magnitude. By calling $K_i(q)$ the "proper solution" corresponding to λ_i, any series

$$R(q, t) \equiv \sum_i A_i e^{-\lambda_i t} K_i(q)$$

satisfies, simultaneously, both (3.2.7) and the conditions at the limits. In addition, it satisfies the initial condition $R(q, 0) \equiv R_0(q)$ if the coefficients A_i are chosen so that $\sum_i A_i K_i(q) \equiv R_0(q)$, i.e., if they are given by the expansion of the function $R_0(q)$ in series of functions $K_i(q)$. We know that such an expansion is possible for a function $R_0(q)$ which is continuous and equal to zero at the limits $q = 0$ and $q = 1$. To express the expansion, it suffices to write equation (3.2.9) in the reduced form

$$\frac{d^2\overline{K}}{dr^2} = \frac{-2\lambda}{w(q)\phi^2(q)} \overline{K},$$

designating the new variable, $\int_0^q \phi(q)\,dq$, by r, which is the function of total probability $\Phi(q)$. We know, then, that the proper solutions $\overline{K}_i(r)$ are orthogonal (and can be taken to be normalized) with respect to the function $1/w\phi^2(q)$, i.e., that

$$\int_0^1 \frac{\overline{K}_i(r)\overline{K}_j(r)}{w\phi^2(q)}\,dr = 0,$$

or, by going back to the variable q,

$$\int_0^1 \frac{K_i(q)K_j(q)}{w\phi(q)}\,dq = 0,$$

where

$$\int_0^1 \frac{[K_i(q)]^2}{w\phi(q)} \, dq = 1.$$

The coefficients A_i of the expansion of $R_0(q)$ are, therefore, of the form

$$A_i = \int_0^1 \frac{R_0(q)K_i(q)}{w\phi(q)} \, dq,$$

which gives the solution to the problem as

$$R(q, t) = \sum_{i=1}^{\infty} A_i e^{-\lambda_i t} K_i(q), \qquad (3.2.10)$$

which is a uniformly converging series. We note that the magnitude of the decrease of the difference $R(q, t)$ between the asymptotic law at instant t and the integral law is on the order $e^{-\lambda_1 t}$, λ_1 being the first proper value, unless in the function $K_1(q)$ the initial deviation, $R_0(q)$ is not orthogonal to $1/w\phi(q)$. The rate of the process is thus characterized.

It is easy to resolve the problem completely in the case previously studied, where $\delta(q)$ can be replaced by the linear function $\delta(q) = -k(q - \bar{q})$. Then equation (3.2.9), where $w = q(1 - q)/2N$ becomes Gauss's equation

$$q(1 - q)K'' + [1 - 2q + 4Nk(q - \bar{q})]K' + 4N\lambda K = 0. \quad (3.2.9')$$

The Gaussian parameters here are α and β, the roots of

$$\alpha^2 + (4Nk - 1)\alpha - 4N\lambda = 0 \qquad \text{and} \qquad \gamma = 1 - 4Nk\bar{q} \quad (3.2.11)$$

Calling $F(\alpha, \beta, \gamma; q)$ the hypergeometric series, the general solution of (3.2.9') is

$$C_1 F(\alpha, \beta, \gamma; q) + C_2 q^{4Nk\bar{q}} F(\alpha', \beta', \gamma'; q),$$

where

$$\alpha' = \alpha + 1 - \gamma, \qquad \beta' = \beta + 1 - \gamma, \qquad \gamma' = 2 - \gamma.$$

The solutions that equal zero for $q = 0$ correspond to $C_1 = 0$. There will be, therefore, "proper solutions" becoming equal to zero either when $q = 0$ or when $q = 1$, provided that $F(\alpha', \beta', \gamma'; 1)$, which, according to Gauss's theory of equations, is equal to $\dfrac{\Gamma(\gamma')\Gamma(\gamma' - \alpha' - \beta')}{\Gamma(\gamma' - \alpha')\Gamma(\gamma' - \beta')}$, equals 0; this requires that α or β be equal to a whole number, $n \geqslant 1$, i.e., that equation (3.2.11) have a whole, positive root n which gives for λ the "proper values" $\lambda_n = n^2/4N + n(k - 1/4N)$, values that increase from k to $+\infty$.

The corresponding proper standardized solutions are the hypergeometric functions

$$K_n(q) = h_n F(n + 4Nk\bar{q}, 1 - n - 4Nk + 4Nk\bar{q}, 1 + 4Nk\bar{q}; q).$$

The constants h_n are chosen to give

$$\int_0^1 \frac{[K_n(q)]^2}{q^{4Nk\bar{q}}(1 - q)^{4Nk(1 - \bar{q})}} \, dq = 1.$$

The coefficients A_n are given by

$$A_n = \int_0^1 \frac{R_0(q)K_n(q)}{q^{4Nk\bar{q}}(1 - q)^{4Nk(1 - \bar{q})}} \, dq.$$

The difference is given by the formula (3.2.10).

Since $\lambda_1 = k$, the order of magnitude of the decrease of this difference will be, in general, that of e^{-kt}; the number t of generations needed to approach the state of asymptotic equilibrium appreciably, therefore, will be on the order of magnitude of $1/k$. We have seen [§3.2.1(A)] that when $\delta(q)$ has the general form derived at the end of §3.2(E), but the distribution remains, over time, sufficiently concentrated around the value $\bar{q}$, we take

$$k = -\delta'(\bar{q}) = u + v - (1 - 2\bar{q})t - w\bar{q}(2 - 3\bar{q});$$

k is, then, on the order of magnitude of the largest (in absolute value) of the quantities u, v, t, w. When all these quantities are small, $1/k$ is large, and the number of generations needed to approach equilibrium is considerable. We cannot assume, therefore, that a natural population has reached the state of equilibrium unless conditions have remained the same during a very long period of time.

The preceding method does not apply any longer in cases where there are neither mutations nor migrations, i.e., when $u = v = 0$, because then $K = 0$ and the density of asymptotic probability, $\phi(q) \, dq$, equals zero at any point between 0 and 1. All probability is concentrated at the two extremes, $q = 0$ and $q = 1$. The manner in which this asymptotic state is reached can be studied by a different method [11].

3.3 INFLUENCE OF MIGRATION

The hypothesis by which Wright [22, 23, 24] explains the effects of migration would apply well only to an island population receiving migrants from a large continental population with constant composition. A scheme closer to the actual situation, which takes into account the interaction of one group with another by migration, would be the following. Let a population be distributed over an area A with a density $\delta(P)$ at point P with coordinates (x, y). Let us assume that each individual, from the time of birth to the reproductive stage, has a known probability, $f(P, Q) \, dS_Q$, of migrating from the point P to an elementary area, dS_Q, centered at point Q $\left(\iint_A f(P, Q) \, dS_Q = 1 \right)$. According to Bayes's formula, each parent of an individual born at point Q will have the known probability,

$$g(P, Q) \, dS_P = \delta(P) f(P, Q) \, dS_P \bigg/ \iint_A \delta(P) f(P, Q) \, dS_P,$$

of being born in an area dS_P centered around point P $\left(\iint_A g(P, Q) \, dS_P = 1 \right)$.

Let $X_n(C)$ be the random variable representing the state of a Mendelian locus in an individual of the nth generation born at point C. *A priori*, $X_n(C)$ will take the values 1 or 0, corresponding to the allelic states a or A, with *a priori* probabilities q and $p = 1 - q$, depending on the point C and the rank n of the generation; the X_ns relative to two different points C will have a stochastic relation.* The random variables $X_{n+1}(D)$ relative to the following generation will have conditional probabilities well-determined on the basis of the $X_n(C)$ values. According to the theory of Markov chains, it follows that the *a priori* probabilities of the $X_n(C)$s and their relationships will tend eventually toward a stationary state, independent of the rank, n, of the generation. It is this stationary state we propose to study.

If u and v are the probabilities of mutation of a into A and of A into a in each generation, the conditional expectation of the random variable X' relative to a locus of an offspring of a specified parent will be

$$\mathfrak{M}(X') = (1 - u)X + v(1 - X),$$

X being the specified value of the random variable attached to the corresponding locus in the parent. This can be written

$$\mathfrak{M}(X') = (1 - k)X + kc,$$

calling c the quantity $\bar{q} = v/(u + v)$ and k the quantity $v + u$ corresponding to the mutation pressure.† Since there is no stochastic relation among children other than the one resulting from the eventual relation among their parents, the joint moments $\mathfrak{M}[X'(C)X'(D)]$

---

* If the coefficient of coancestry between individuals located in places C and D is called $\phi_n(C, D)$, the random variables $X_n(C)$ and $X_n(D)$ have an *a priori* probability of ϕ_n of being identical and a probability of $1 - \phi_n$ of being stochastically independent; this gives, as the value of their *a priori* correlation coefficient, $\phi_n(C, D)$. The asymptotic value, $\phi(C, D)$, of this coefficient will be calculated further; it is useful to know that it is *the same as the coefficient of coancestry between C and D.*

† If there is a constant selection pressure in favor of the heterozygote, we know that it will be expressed, approximately, for a large number of individuals by a formula of the same form (k and c naturally having other values). The calculations we shall perform will, therefore, be a first approximation applicable to constant selection, but they exclude, naturally, the "geographic selection" that depends on location.

of a certain number of random variables X' of the $(n + 1)$th generation will be linear combinations of the products of the variables X of the parents, if the latter are known; if they are unknown, the X' will be linear combinations of the mathematical expectations of these products, i.e., of the joint moments of the nth generation. By equating the joint moments of the two generations, we shall obtain linear integral equations for determining these moments. We shall indicate only the calculations for the moments of orders 1 and 2.

The mathematical expectation, $\mathfrak{M}(Q)$, of $X(Q)$ will be given by

$$\mathfrak{M}[X(Q)] = \iint_A \{(1 - k)\mathfrak{M}[X(P)] + kc\}g(P, Q)\, dS_P,$$

that is,

$$\mathfrak{M}(Q) = \iint_A (1 - k)\mathfrak{M}(P)g(P, Q)\, dS_P + kc,$$

an equation whose only solution, if $k = (u + v) > 0$, is

$$\mathfrak{M}(P) = \text{constant} = c = \frac{v}{u + v}.$$

The mathematical expectation is therefore independent of the geographical position. In the calculations that follow $X - c = Y$, $\mathfrak{M}(Y) = 0$, and from one generation to the next $\mathfrak{M}(Y') = (1 - k)Y$.

The variance of X, or of Y, will be

$$s^2 = \mathfrak{M}(Y^2) = \mathfrak{M}(X^2) - c^2 = c(1 - c).$$

The joint first moment of the two random variables $Y(C)$ and $Y(D)$ of the same generation will be designated by $\mathfrak{M}[Y(C)Y(D)] = s^2\phi(C, D)$; $\phi(C, D)$ is both coefficient of coancestry and *a priori* correlation coefficient of these two random variables and also of $X(C)$ and $X(D)$.* Let us call $\phi(C, C)$ its limit, obviously less than one, when D gets infinitely close to C, the two loci remaining

---

* Also of the local frequencies q_C and q_D in places C and D, because these are local arithmetic means of such random variables.

distinct (they may be, in case of random mating, the two homologous loci of the same individual).

Two loci, $Y_{n+1}(C)$ and $Y_{n+1}(D)$, of two individuals in the $(n + 1)$th generation born in C and D will have the probability $g(E, C)g(F, D) dS_E dS_F$ of coming from parents born in E and F and the probability $g(E, C)g(F, D) dS_E^2$ of coming from parents both born in the same neighborhood of a single site E; in the latter case they will have the conditional probability $1/[2\delta(E) dS_E]$ of coming from the same locus of the same parent and the probability $1 - 1/[2\delta(E) dS_E]$ of coming from loci infinitely close but distinct.* We have, therefore, when the places of birth, E and F, of the parents are known (conditional expectation),

$$\mathfrak{M}_n[Y_{n+1}(C)Y_{n+1}(D)] = (1 - k)^2 Y_n(E)Y_n(F);$$

and when they are unknown (*a priori* expectation),

$$\mathfrak{M}[Y_{n+1}(C)Y_{n+1}(D)]$$
$$= \mathfrak{M}\{\mathfrak{M}_n[Y_{n+1}(C)Y_{n+1}(D)]\}$$
$$= (1 - k)^2 \iint_A \iint_A \mathfrak{M}[Y_n(E)Y_n(F)]g(E, C)g(F, D) dS_E dS_F.$$

$\mathfrak{M}[Y_n(E)Y_n(F)]$ should be taken as equal to $s^2\phi_n(E, F)$ if the elements of area dS_E and dS_F are distinct; if they are not distinct, $\mathfrak{M}[Y_n(E)Y_n(F)]$ should be taken as equal to

$$\frac{\mathfrak{M} Y_n(E)^2}{2\delta(E) dS_E} + [1 - 1/2\delta(E) dS_E]s^2\phi_n(E, E)$$

that is, equal to

$$s^2\phi_n(E, E) + \frac{s^2[1 - \phi_n(E, E)]}{2\delta(E) dS_E};$$

dividing by s^2 we have the "Fredholm iteration":

---

* Formula for monoecious random mating; in case of separate sexes, $\delta(E)$ is twice the harmonic mean of male and female densities in E.

$$\phi_{n+1}(C, D) = (1 - k)^2 \iint_A \iint_A \phi_n(E, F)g(E, C)g(F, D) \, dS_E \, dS_F$$

$$+ (1 - k)^2 \iint_A \frac{1 - \phi_n(E, E)}{2\delta(E)} g(E, C)g(E, D) \, dS_E.$$

$$(3.3.1)$$

In the stationary state, if $\phi(E, E) = \lim \phi_n(E, E)$ was a known function, (3.3.1) would be, for the unknown function $\phi(C, D) = \lim \phi_{n+1}(C, D)$, a Fredholm equation with an integrable kernel of norm $(1 - k)^2 < 1$ (if $k > 0$); it would then have a unique solution given, whatever the initial values, by the same integration as for zero initial values:

$$\phi(C, D) = \iint_A \frac{1 - \phi(E, E)}{2\delta(E)} \sum_{n=0}^{\infty} (1 - k)^{2n+2} g_n(E, C)g_n(E, D) \, dS_E,$$

$$(3.3.2)$$

by setting

$$g_1(E, C) = \iint_A g(E, F)g(F, C) \, dS_F,$$

$$g_n(E, C) = \iint_A g_{n-1}(E, F)g(F, C) \, dS_F.$$

By taking $E \equiv C$, we obtain a second Fredholm equation for the determination of $\phi(E, E)$:

$$\phi(C, C) = \iint_A \frac{1 - \phi(E, E)}{2\delta(E)} \sum_{n=0}^{\infty} (1 - k)^{2n+2} g_n^2(E, C) \, dS_E. \quad (3.3.3)$$

This equation in general (when its kernel is integrable and of norm < 1) has a single solution, $\phi(E, E)$; by putting it into (3.3.2), we obtain $\phi(C, D)$.

REMARK I. A PARTIAL DIFFERENTIAL EQUATION
APPROXIMATING (3.3.1).

We may introduce the moments of the migration law, i.e.,

$$m_{pq} = \iint (x_E - x_C)^p (y_E - y_C)^q g(E, C) \, dS_E,$$

by replacing, in the second term of (3.3.1), $\phi_n(E, F)$ by its Taylor development,

$$\phi_n(E, F) = \phi_n(C, D) + (\vec{CE}\cdot\vec{\nabla}_C + \vec{DF}\cdot\vec{\nabla}_F)\phi_n(C, D) + \ldots$$
$$+ \frac{1}{p!}(\vec{CE}\cdot\vec{\nabla}_C + \vec{DF}\cdot\vec{\nabla}_F)^p\phi_n(C, D) + \ldots,$$

using the symbol $\vec{CE}\cdot\vec{\nabla}_C$ for the operator

$$(x_E - x_C)\frac{\partial}{\partial x_C} + (y_E - y_C)\frac{\partial}{\partial x_C},$$

its powers being defined as usual $\left[\dfrac{\partial^{p+q}}{\partial x_C^p \, \partial y_C^q} = \left(\dfrac{\partial}{\partial x_C}\right)^p\left(\dfrac{\partial}{\partial y_C}\right)^q\right]$.

It is now easy to express the double area integral in the second term of (3.3.1) as a function of the partial derivatives of $\phi_n(C, D)$, the coefficients being the moments, m_{pq}, calculated from place C, and the similar moments, m'_{pq}, calculated from place D; the beginning of this integral is (considering a symmetrical case, for the sake of simplicity, because the odd moments are then equal to zero):

$$\phi_n(C, D) + \frac{1}{2}\left[m_{20}\frac{\partial^2\phi_n}{\partial x_C^2} + 2m_{11}\frac{\partial^2\phi_n}{\partial x_C \, \partial y_C} + m_{02}\frac{\partial^2\phi_n}{\partial y_C^2}\right.$$
$$\left. + m'_{20}\frac{\partial^2\phi_n}{\partial x_D^2} + 2m'_{11}\frac{\partial^2\phi_n}{\partial x_D \, \partial y_D} + m'_{02}\frac{\partial^2\phi_n}{\partial y_D^2} + \ldots\right].$$

(The formula for unidimensional or tridimensional cases is naturally of the same form.)

If the moments and their products are negligible from some order, and if we replace $\phi_n(C, D)$ and $\phi_{n+1}(C, D)$ by their equilibrium expression, $\phi(C, D)$, this last function is a solution (which tends to zero when distance CD tends to infinity) of a linear partial differential equation, of which the nonhomogeneous term $\displaystyle\iint_A \frac{1 - \phi(E, E)}{2\delta(E)}g(E, C)g(E, D)\,dS_E$ itself tends to zero when CD tends to infinity.

3.3.1 Special Case of "Homogeneous and Isotropic" Migration

Let us suppose that the area occupied by the population can be considered unlimited, that the density $\delta(E)$ is constant (in space and time) and that $f(P, Q)$ depends only on the distance $PQ = r$; then

$g(P, Q)$ is equal to $f(P, Q)$. Let us set $g(P, Q) = g(r)$, so that it becomes a function of a single variable, no longer of four; similarly $g_n(P, Q) = g_n(r)$. From (3.3.3) we get

$$\phi(C, C) = \iint_A \frac{1 - \phi(E, E)}{2\delta} \sum_{n=0}^{\infty} (1 - k)^{2n+2} g_n^2(EC) \, dS_E,$$

an integral equation whose solution by successive approximations gives $\phi(C, C) = \text{constant} = \phi_0$. It follows that

$$\phi_0 = H \frac{1 - \phi_0}{2\delta},$$

from which

$$\phi_0 = \frac{H}{2\delta + H}, \qquad (3.3.3')$$

where

$$H = \iint_A \sum_{n=0}^{\infty} (1 - k)^{2n+2} g_n^2(r) \, dS.$$

This is the value of the correlation coefficient between two closely located loci.

Equation (3.3.2) shows that $\phi(C, D)$ depends only on the distance CD. Let $\phi = \phi(CD)$; then

$$\phi(CD) = \frac{1 - \phi_0}{2\delta} \iint_A \sum_{n=0}^{\infty} (1 - k)^{2n+2} g_n(EC) g_n(ED) \, dS_E. \quad (3.3.4)$$

This is the correlation coefficient between two loci whose distance is CD.* We can express, in algebraic terms, the "products of composition" (or "convolutions") which appear in (3.3.4) by considering the Fourier transforms,

$$F(u, v) = \iint e^{iux+ivy} g(\sqrt{x^2 + y^2}) \, dx \, dy$$

and

---

* Or the coefficient of coancestry between C and D (see p. 65n). For random mating, $\phi(C, C) = \phi_0$ (coancestry between closely located loci) is also the inbreeding coefficient (coancestry between the two homologous loci of one individual).

$$K(u, v) = \iint e^{iux+ivy}\phi(\sqrt{x^2 + y^2})\, dx\, dy,$$

because we have

$$K(u, v) = \frac{1 - \phi_0}{2\delta}\left[\sum_{p=1}^{\infty} (1 - k)^{2p}F^{2p}\right]$$

$$= \frac{1 - \phi_0}{2\delta}(1 - k)^2 F^2/[1 - (1 - k)^2 F^2],$$

a formula which is also obtained by applying the Fourier transform directly to formula (3.3.1).

Thus K is expressed as a function of $F(u, v)$, which is known. From this, by inversion of the Fourier transform with two variables, we have

$$\phi = \frac{1 - \phi_0}{8\pi^2\delta}\iint e^{-iux-ivy}\frac{(1 - k)^2 F^2}{1 - (1 - k)^2 F^2}\, du\, dv. \qquad (3.3.5)$$

(By setting $x = y = 0$, we find again the linear equation for ϕ_0.)

These calculations can be carried still further by assuming that the displacement of each individual is a random movement following the scheme of Polya, i.e., that the law $f(r)\, dS = g(r)\, dS$ is an isotropic normal law,

$$G(\sigma^2)\, dS = (1/2\pi\sigma^2)e^{-r^2/2\sigma^2}\, dS,$$

which gives

$$F(u, v) = e^{-(\sigma^2/2)(u^2+v^2)},$$

$$K(u, v) = \frac{1 - \phi_0}{2\delta}\sum_{p=1}^{\infty} (1 - k)^{2p}e^{-(2p\sigma^2/2)(u^2+v^2)}.$$

From this we deduce series (3.3.4), which is easier to calculate than (3.3.5), and from which

$$\phi(r) = \frac{1 - \phi_0}{2\delta}\sum_{p=1}^{\infty} (1 - k)^{2p}G(2p\sigma^2). \qquad (3.3.4')$$

This series is uniformly convergent when $k > 0$, since

$$G(2p\sigma^2) \leqslant \frac{1}{4\pi p\sigma^2}.$$

Formula (3.3.3) can be found here by making $r = 0$, which leads us to calculate

$$H = \sum_{p=1}^{\infty} (1 - k)^{2p}[1/4\pi p\sigma^2]$$

$$= [1/4\pi\sigma^2] \int_0^{(1-k)^2} \left[\sum_{p=1}^{\infty} x^{p-1} \right] dx$$

$$= -\log [1 - (1 - k)^2]/4\pi\sigma^2 = -\log (2k - k^2)/4\pi\sigma^2,$$

from which

$$\phi_0 = 1/[1 - 8\pi\sigma^2\delta/\log (2k - k^2)]. \qquad (3.3.3'')$$

We can calculate ϕ_0 easily, from the pressure k (of overdominance or of mutation) and from the number $\pi\sigma^2\delta$ of individuals in a circle of radius σ, in which resides, on the average, 40 per cent of the individuals born at its center); the smaller these two quantities are, the closer ϕ_0 is to 1 (local quasihomogeneity); next, we deduce from (3.3.4'),

$$\frac{\phi(r)}{\phi_0} = \frac{\sum_{p=1}^{\infty} (1 - k)^{2p} \left(\frac{1}{4\pi p\sigma^2} \right) e^{-r^2/4p\sigma^2}}{\sum_{p=1}^{\infty} (1 - k)^{2p} \left(\frac{1}{4\pi p\sigma^2} \right)}, \qquad (3.3.4'')$$

which shows that the correlation to the distance r decreases from ϕ_0 to 0 when r increases from 0 to ∞. The numerical value of this ratio depends only on two quantities, k and r/σ; it is, therefore, easy to set up tables that will enable us to interpret the experimental results with the help of this formula.

REMARK II.

To calculate (3.3.4'') numerically, we can develop the numerator according to the powers of r^2, arriving at the series

$$\sum_{p=1}^{\infty} (1 - k)^{2p}/p^n,$$

whose sum is

$$\int_0^{(1-k)^2} \frac{\log [(1 - k^2)/X]^{n-1}}{(n - 1)!} \frac{dX}{1 - X};$$

we also deduce from this that the numerator of (3.3.4″) is equal to

$$-\frac{1}{4\pi\sigma^2} \int_0^{\log (1 - (1 - k)^2)} J_0 \left[\frac{r}{\sigma} \sqrt{\log \frac{(1 - k)^2}{1 - e^u}} \right] du,$$

J_0 being the Bessel function. By letting $r = 0$, we find again the denominator H.

REMARK III.

If k tends toward zero, the numerator and the denominator of (3.3.4″) tend toward infinity, but their difference remains finite (according to the properties of J_0); therefore, $H \longrightarrow \infty$, ϕ_0 and $\phi \longrightarrow 1$; and the population tends toward complete homogeneity, which is inevitable in any population with a finite size in the absence of mutations.

REMARK IV.

We may, in the partial differential equation shown to approximate (3.3.1), when σ is small with respect to $r\sqrt{k}$, keep only the second moments $m_{20} = m_{02} = m'_{02} = m'_{20} = \sigma^2$, $m_{11} = m'_{11} = 0$ (the higher moments, being higher powers of σ, give negligible characteristic roots); r being large with respect to σ, $\iint_A g(E, C)g(E, D)$ is negligible and $\phi(C, D) = \phi(r)$ is a solution, null at infinity, of the homogeneous Helmholtz equation

$$\phi(r) = (1 - k)^2[\phi(r) + \sigma^2 \Delta\phi(r)],$$

$\Delta\phi$ being the "Laplacian" $\dfrac{\partial^2\phi}{\partial x^2} + \dfrac{\partial^2\phi}{\partial y^2}$, which, in polar coordinates r and θ, is, like $\phi(r)$, independent of θ and equal to $\dfrac{\partial^2\phi}{\partial r^2} + \dfrac{1}{r}\dfrac{\partial\phi}{\partial r}$; so (when neglecting k^2) we obtain the Bessel equation

$$\frac{\partial^2\phi}{\partial r^2} + \frac{1}{r}\frac{\partial\phi}{\partial r} - \frac{2k}{\sigma^2}\phi = 0.$$

Of the two distinct solutions, I_0 and K_0, only K_0 is bounded, thus giving the correlation (or coefficient of coancestry):

$$\phi(r) = aK_0\left(r\frac{\sqrt{2k}}{\sigma}\right) \sim \frac{a}{\sqrt{r}}e^{-\sqrt{2kr/\sigma}},$$

where a is a constant and r is much greater than σ.

The same equation, and the same result, is true for every migration law all of whose reduced moments are bounded [15], and the Helmholtz equation is valid for an isotropic migration of any dimensionality. So, in unidimensional cases, $\dfrac{\partial^2\phi}{\partial r^2} - \dfrac{2k}{\sigma^2}\phi = 0$ gives an exponential decrease proportional to $\exp -\sqrt{2k}r/\sigma$. This exponential decrease has also been found in discontinuous cases [13, 15]. Weiss and Kimura [25] extended the formulas to the tridimensional case; $\Delta\phi - \dfrac{2k}{\sigma^2}\phi = 0$ gives $\dfrac{\partial^2\phi}{\partial r^2} + \dfrac{2}{r}\dfrac{\partial\phi}{\partial r} - \dfrac{2k}{\sigma^2}\phi = 0$, giving a decrease proportional to $\dfrac{1}{r}\exp -\sqrt{2k}r/\sigma$.

In all these formulas, σ is the standard deviation of the migration along each axis of coordinates (migration may not be normal); the correlation $\phi(r)$ with large distance r depends only on the ratio r/σ and on the rate k.

These asymptotic formulas (3.3.4 and its varieties) are independent of density, δ, so we can use them in a population whose density varies considerably over the years (as with Chetverikov's waves of vitality).

If the individuals show a tendency to stay grouped in "colonies" or "swarms," we shall take that into account by postulating that each individual has respective probabilities α and $(1 - \alpha)$ of making an infinitely small displacement with variance ϵ^2 or of making a migration to a distant point with variance σ^2, i.e., by taking $g(r) = \alpha G(\epsilon^2) + (1 - \alpha)G(\sigma^2)$, which gives

$$F(u, v) = \alpha e^{-(\epsilon^2/2)(u^2+v^2)} + (1 - \alpha)e^{-(\sigma^2/2)(u^2+v^2)},$$

from which we have a formula for $\phi(r)$, i.e., the same asymptotic expression, but with variance $\alpha\epsilon^2 + (1 - \alpha)\sigma^2$.

The experimental determination of the correlation as a function of distance, for verification of this theory, can be done in several different ways.

(1) We can measure the frequency, q_i, of a Mendelian gene (without geographic selection) at a large number of points, P_i, of a wide territory; we shall take the general mean of these frequencies as an estimate of c, and the mean of all the quantities $\dfrac{(q_i - c)(q_j - c)}{c(1 - c)}$, calculated from two points, P_i and P_j, whose distance is r, as an estimate of $\phi(r)$ and of its decrease when r increases. This verification was made by Lamotte on *Cepaea nemoralis* [**10, 16**].

(2) We can measure on different individuals a biometrical trait, neutral for selection, whose intensity can be considered the additive effect of a certain number of independent Mendelian random variables, X_i with expectations M_i (each X_i assuming values s_i and t_i with probabilities q_i and p_i as functions of the location); the mean correlation between two individuals, I and I', situated at a distance r will be an estimate of

$$\frac{\mathfrak{M}[\Sigma(X_i - M_i)\Sigma(X_i' - M_i)]}{\mathfrak{M}[\Sigma(X_i - M_i)]^2} = \frac{\Sigma\mathfrak{M}[(X_i - M_i)(X_i' - M_i)]}{\Sigma\mathfrak{M}(X_i - M_i)^2},$$

that is, will be an estimate of $\phi(r)$ if we postulate that the rate of mutation, k, is the same for all the genes concerned.* We must remember, however, that the correlation decreases if a fraction of the variability is not genetic (we then have to multiply by the "heritability").

3.3.2 Other Applications

(A) Panmixia in a finite isolated population of N individuals can be studied by assuming that the occupied area, A, is equal to 1,

---

* This is only approximate. If we consider the fact that the two random variables X_i of each individual have, with random mating (see p. 70), a coefficient of inbreeding equal to ϕ_0, the numerator and denominator are $4\phi(r)\Sigma'(X_i - M_i)^2$ and $2(1 + \phi_0)\Sigma'(X_i - M_i)^2$, the summation Σ' being now extended only to nonhomologous loci, from which the correlation [13] between the genetic components of metrical traits (without dominance nor epistasis) is $\dfrac{2\phi(r)}{1 + \phi(0)}$.

δ is equal to N, and $g(P, Q)$ is equal either to 1, if P and Q are within A, or to 0, if they are outside. We have, then, for C and D in A,

$$\phi(C, D) = \phi(C, C) = \frac{1 - \phi(E, E)}{2\delta} \sum_{n=0}^{\infty} (1 - k)^{2n},$$

which gives

$$\phi = \text{constant} = \frac{(1 - k)^2}{2N[1 - (1 - k)^2] + (1 - k)^2}.$$

By equating this expression to ϕ_0 given by (3.3.3) or (3.3.3''), we obtain the size N of a panmictic group "equivalent" to a group occupying a very small area and constituting part of a population with random isotropic migration. Let

$$N[(1 - k)^{-2} - 1] = \frac{4\pi\sigma^2\delta}{-\log(2k - k^2)}.$$

This concept of "equivalent effective number" introduced by Wright [22, 23, 24], following entirely different reasoning, does not have the weight that he attributes to it, because it does not account for the correlation with distance.

(B) We can try to formulate a scheme of homogeneous but non-isotropic migration (in an unlimited population of constant density) by postulating that the displacement of an individual results from two independent displacements with different laws of probability, in two rectangular directions, i.e., that

$$g(P, Q)\, dS_p = m(x_1 - x_2)n(y_1 - y_2)\, dx_1\, dy_1.$$

Designating the coordinates of P and Q by x_1, y_1, x_2, y_2 (m and n being two functions each of one variable, whose integral from $-\infty$ to $+\infty$ is equal to 1), and setting

$$m_{p+1}(x_1 - x_2) = \int_{-\infty}^{+\infty} m_p(x_1 - x_3)m(x_3 - x_2)\, dx_3$$

(and similarly for n), formula (3.3.2) becomes

$$\phi(P, Q) = \iint \frac{1 - \phi(E, E)}{2\delta} \left[\sum_{p=0}^{\infty} (1 - k)^{2p+2} m_p(x_3 - x_1) m_p(x_3 - x_2) \right.$$

$$\left. n_p(y_3 - y_1) n_p(y_3 - y_2) \right] dx_3 \, dy_3.$$

In particular, if we take for $g(P, Q)$ Gauss's nonisotropic law,

$$g(P, Q) = \frac{1}{2\pi\sigma\rho} e^{-\frac{(x_2 - x_1)^2}{2\sigma^2} - \frac{(y_2 - y_1)^2}{2\rho^2}},$$

we find $\phi(E, E) = $ constant $= \phi_0$, and we have

$$\phi(P, Q) = \frac{1 - \phi_0}{4\pi\delta\sigma\rho} \sum_{p=1}^{\infty} \frac{(1 - k)^{2p}}{2p} e^{-\frac{(x_2 - x_1)^2}{2p\sigma^2} - \frac{(y_2 - y_1)^2}{2p\rho^2}},$$

from which we calculate ϕ_0 by making $x_2 = x_1$ and $y_2 = y_1$. For long distances, we obtain a homogeneous partial differential equation (of elliptic type). We could introduce an analogous scheme with three dimensions to represent the variability of an aquatic population according to the two coordinates of surface and depth. The approximate partial differential equation is easy to write, since it is a generalization of the Helmholtz equation on p. 73.

3.4 APPENDIX: DISCONTINUOUS MIGRATIONS

The case of discontinuous migrations refers to the model in which the individuals of each generation do not inhabit a continuous area but rather a discrete set of places (still called A), each place being looked at as a point (still called C or D for two offsprings, I and J, taken in F_{n+1}, and E or F for the possible places of their parents, P_I and P_J). The integrations in formula (3.3.1) have to be replaced by summations on the lattice A of all possible places E and F; $g(E, C)$ is the probability, in place C, of an individual coming from place E. We have

$$\sum_{E \in A} g(E, C) = 1;$$

$\dfrac{1}{2\delta(E)\,dS_E}$ has to be replaced by the conditional probability for two parents, successively taken in place E, to be identical, i.e., by $1/2N_E$ if there are N_E equally probable monoecious parents:

When there are N_{1E} and N_{2E} equally probable fathers and mothers, the conditional probability is still $1/2N_E$ if we call N_E the double of the harmonic mean, $1/N_E = 1/4(1/N_{1E} + 1/N_{2E})$. Equation (3.3.1) may now be written

$$\phi_{n+1}(C, D) = (1 - k)^2 \sum_{E \in A} \sum_{F \in A} \phi_n(E, F)g(E, C)g(F, D)$$

$$+ (1 - k)^2 \sum_{E \in A} \frac{1 - \phi_n(E, E)}{2N_E} g(E, C)g(E, D). \tag{3.4.1}$$

If we call $\phi(C, D)$ some solution independent of n, the difference $\phi_{n+1}(C, D) - \phi(C, D) = \psi_{n+1}(C, D)$ is related to $\phi_n(E, F) - \phi(E, F) = \psi_n(E, F)$ by a recurrence now homogeneous, and $\sup_{E,F} \psi_n(E, F)$ tends to zero when n goes to infinity; so $\phi(C, D)$ (when it is supposed to exist) is unique, and is the limit of ϕ_n when n goes to infinity. Conversely, a limit existing for some particular initial condition is a solution independent of n, and this is the same as the limit for any initial condition.

So, to obtain the limit, it is sufficient to take the particular initial condition $\phi_0(E, F) \equiv 0$, which gives

$$\phi(C, D) = (1 - k)^2 \sum_{E \in A} \frac{1 - \phi(E, E)}{2N_E} \sum_{n=0}^{\infty} (1 - k)^{2n+2}g_n(E, C)g_n(E, D),$$

$$\tag{3.4.2}$$

putting

$$g_n(E, C) = \sum_{F \in A} g_{n-1}(E, F)g(F, C). \tag{3.4.2'}$$

Let us now suppose that the migration is homogeneous, i.e., that $g(E, C)$ depends only on the components of the vector $\overrightarrow{CE}$, each of which components may be supposed to be integers p and q.* Let us put $g(E, C) = \mu(p, q)$. The migration law may then be defined by the "generating function"

$$G(\alpha, \beta) = \sum_{E \subset A} \mu(p, q)\alpha^p\beta^q,$$

which converges absolutely when $|\alpha| = 1$ and $|\beta| = 1$ (putting $\alpha = e^{i\phi}$ and $\beta = e^{i\psi}$, we have a Fourier series).

In the same manner, we put

$$G_m(\alpha, \beta) = \sum_{E \subset A} \mu_m(p, q)\alpha^p\beta^q, \qquad (3.4.3)$$

where $\mu_m(p, q)$ is $g_m(E, C)$ expressed as a function of the components p and q of the vector $\overrightarrow{CE}$; i.e., by introducing in (3.4.2') the components p' and q' of $\overrightarrow{FE}$ and p_0 and q_0 of $\overrightarrow{CF}$, related by $p = p' + p_0$, $q = q' + q_0$,

$$G_m(\alpha, \beta) = \sum_{E \subset A} g_m(E, C)\alpha^{p'+p_0}\beta^{q'+q_0}$$

$$= \sum_{E \subset A,\ F \subset A} g_{m-1}(E, F)\alpha^{p'}\beta^{q'}g(F, C)\alpha^{p_0}\beta^{q_0}.$$

Since there is absolute convergence, we may make the summations in any order. By summing first with respect to E, since the last three factors do not depend on E, we have

$$G_m(\alpha, \beta) = \sum_{F \subset A} G_{m-1}(\alpha, \beta)g(F, C)\alpha^{p_0}\beta^{q_0};$$

G_{m-1} may be made a factor in the summation, which then gives

$$\sum_F g(F, C)\alpha^{p_0}\beta^{q_0} = G(\alpha, \beta),$$

so we have

$$G_m(\alpha, \beta) = G_{m-1}(\alpha, \beta)G(\alpha, \beta),$$

---

* If the notations used for each point consist of two integers ("integer coordinates"), p and q are the excesses of the coordinates of E over the coordinates of C.

and by iteration,

$$G_m(\alpha, \beta) = [G(\alpha, \beta)]^m.$$

Equation (3.4.2) for $\phi(C, D)$ may then be transformed into an equation giving a "generating function" of ϕ; it is sufficient to note that if we complete the definition of homogeneous migration by putting $N_E = N$ *independent of place* E, $\phi(E, E) = $ constant gives a solution for $\phi(C, D)$, a solution which is known to be unique; so we may put $\phi(E, E) = $ constant $= \phi_0$.* The right-hand side of (3.4.2), like the left-hand side, depends only on the components of $\overrightarrow{CD}$, which may be called x and y; $\phi(C, D)$ will now be called $\phi(x, y)$.†

If we multiply (3.4.2) by $\alpha^x\beta^y$, this amounts to multiplying each term of the sum $\underset{E}{\Sigma}$ by $\alpha^p\beta^q\alpha^{x-p}\beta^{y-q}$, p and q being the components of $\overrightarrow{CE}$, $x - p$ and $y - q$ being the components of $\overrightarrow{CD} - \overrightarrow{CE} = -\overrightarrow{DE}$.

We then have

$$\alpha^x\beta^y\phi(x, y) = \frac{(1 - \phi_0)}{2N} \sum_{m=0}^{\infty} (1 - k)^{2m+2}$$

$$\underset{E}{\Sigma} \alpha^p\beta^q g_m(E, C)\alpha^{x-p}\beta^{y-q}g_m(E, D), \tag{3.4.4}$$

setting $g = g_0$.

The right-hand side may, if $k > 0$, be summed up over all values of x and y (components of $\overrightarrow{CD}$), i.e., over all points D, when $|\alpha| = |\beta| = 1$; it is a multiple series whose general term (indexed by m, E, D) has a modulus bounded by $(1 - k)^{2m+2}g_m(E, C)g_m(E, D)$; but because of homogeneity, $\underset{D}{\Sigma} g_m(E, D)$ is the same as $\underset{E}{\Sigma} g_m(E, D)$, and thus equal to 1; thus $\underset{E}{\Sigma} g_m(E, D) \underset{D}{\Sigma} g_m(E, D) = 1$, and then the

---

* We know that this is the inbreeding coefficient at any place.
† We keep the same name for the function of the new variables; they are scalars, not points, and should not be confused.

series is, as $\Sigma (1 - k)^{2m+2}$, absolutely convergent; so we may put
(when $|\alpha| = |\beta| = 1$)

$$\Phi(\alpha, \beta) = \sum_{x,y} \alpha^x \beta^y \phi(x, y).$$

When summing up the right-hand side of (3.3.4), we may begin by noticing that $g_m(E, D) = \mu_m(p - x, q - y)$, and calculating

$$\sum_{x,y} \alpha^{x-p} \beta^{y-q} \mu_m(p - x, q - y) = G_m(1/\alpha, 1/\beta) = [G(1/\alpha, 1/\beta)].$$

Afterwards, the summation over E gives a factor $G_m(\alpha, \beta) = [G(\alpha, \beta)]^m$, and the summation over m gives a geometric series, which gives the same formula as that obtained for the Fourier transform in the continuous case (but extended now to nonsymmetrical migration),

$$\Phi(\alpha, \beta) = \frac{(1 - k)^2[(1 - \phi_0)/2N]G(\alpha, \beta)G(1/\alpha, 1/\beta)}{1 - (1 - k)^2 G(\alpha, \beta)G(1/\alpha, 1/\beta)}.$$

In the symmetrical case, where $g(E, C) = g(C, E)$, we have $G(1/\alpha, 1/\beta) \equiv G(\alpha, \beta)$.

But the "inversion," i.e., the problem of going back from the Fourier series $\Phi(\alpha, \beta)$ to its coefficients $\phi(x, y)$ may be simpler than using integral formulations of these coefficients [formula (3.3.5) written with $e^{iux} = \alpha$, $e^{ivy} = \beta$ and integrated over $x \in (0, 2\pi)$ and $y \in (0, 2\pi)$]. For instance, let us study the *unidimensional* case, when the coefficient of coancestry, for algebraic distance x, is called $\phi(x)$ and has a generating function, $\Phi(\alpha) = \sum_x \alpha^x \phi(x)$, given by

$$\Phi(\alpha) = \frac{(1 - k)^2[(1 - \phi_0)/2N]G(\alpha)G(1/\alpha)}{1 - (1 - k)^2 G(\alpha)G(1/\alpha)}.$$

In most cases where $G(\alpha)$ is a polynomial—as in the case of migration between adjacent groups only, where $G(\alpha) = 1 - 2m + m\alpha + m/\alpha$—we shall see that the expansion into partial fractions gives only two terms with large residues, i.e., those corresponding to the two solutions near 1 of the equation $G(\alpha)G\left(\dfrac{1}{\alpha}\right) = \dfrac{1}{(1-k)^2}$. These two solutions, α_1 and α_2, are obtained by developing $G(\alpha) = \sum\limits_{E} \mu(p)\alpha^p$ [$\overline{CE} = p$ and $\mu(p) = g(E, C)$] into the moments m_p of the migration law, using formulas

$$G(1) = \sum_{E} \mu(p) = 1,$$

$$\left(\frac{\partial G}{\partial \alpha}\right)_{\alpha=1} = \sum_{E} p\mu(p) = m_1,$$

$$\left[\frac{\partial^2 G}{\partial \alpha^2}\right]_{\alpha=1} = \sum_{E} p(p-1)\mu(p) = m_2 - m_1,$$

from which

$$G(\alpha) = 1 + m_1(\alpha - 1) + (m_2 - m_1)(\alpha - 1)^2/2 + 0[(\alpha - 1)^3],$$

$$G(1/\alpha) = 1 - m_1(\alpha - 1)/\alpha + (m_2 - m_1)(\alpha - 1)^2/2\alpha^2$$
$$+ 0[(\alpha - 1)^3]$$
$$= 1 - m_1(\alpha - 1) + m_1(\alpha - 1)^2 + (m_2 - m_1)(\alpha - 1)^2/2$$
$$+ 0[(\alpha - 1)^3],$$

$$G(\alpha)G(1/\alpha) = 1 + (m_2 - m_1^2)(\alpha - 1)^2 + 0[(\alpha - 1)^3],$$

which introduces only (even in the nonsymmetrical case) the variance $\sigma^2 = m_2 - m_1^2$; so we obtain, when k is small and when we equate $G(\alpha)G(1/\alpha)$ to $1/(1 - k)^2 \sim 1 + 2k$, two solutions near 1, α_1 and α_2, given by

$$\alpha_i - 1 = \pm\sqrt{2k/\sigma^2}.$$

Let us recall that the expansion of $\Phi(\alpha)$ into partial fractions uses *all* roots α_i of its denominator* and is $\sum\limits_{i} \dfrac{A_i}{\alpha - \alpha_i}$, each A_i being given by

---

* That is, *all* values α_i such that $G(\alpha_i)G(1/\alpha_i) = 1/(1 - k)^2 \sim 1 + 2k$.

$$A_i = \frac{(1 - k)^2[(1 - \phi_0)/2N]G(\alpha_i)G(1/\alpha_i)}{-(1 - k)^2 \frac{\partial}{\partial \alpha}[G(\alpha)G(1/\alpha)]\alpha}$$

(where the denominator equals α_i);

$$A_i \sim \frac{(1 - \phi_0)/2N}{\sigma^2 \frac{\partial}{\partial \alpha}(\alpha - 1)^2\alpha} = \frac{1 - \phi_0}{4N\sigma^2(1 - \alpha_i)}$$

(where the denominator of the lefthand fractions equals α_i).

This shows that the residues A_1 and A_2 corresponding to the two roots (α_1 and α_2) near 1 are much larger than the others: if we suppose $\alpha_1 < 1 < \alpha_2$, the only terms with negative exponents in the expansion of $\frac{A_1}{\alpha - \alpha_1} + \frac{A_2}{\alpha - \alpha_2}$ will be those* of $\frac{A_1}{\alpha - \alpha_1} = A_1 \sum_{x=0}^{+\infty} \alpha_1^x\alpha^{-x-1}$, thus giving, for negative values $-x$,

$$\phi(-x - 1) \sim A_1\alpha_1^x \sim A_1(1 - \sqrt{2k/\sigma^2})^x \sim A_1e^{-\sqrt{2k}\frac{x}{\sigma}},$$

with

$$A_1 \sim (1 - \phi_0)/[2N\sigma^2(1 - \alpha_1)] \sim (1 - \phi_0)/4N\sigma\sqrt{2k}$$

(and $x > 0$). Similarly, we obtain positive exponents by expansion of:

$$\frac{A_2}{\alpha - \alpha_2} = (-A_2/\alpha_2) \sum_{x=0}^{+\infty} \alpha_2^{-x}\alpha^x$$

which gives (when $x \geqslant 0$)

$$\phi(x) \sim -A_2\alpha_2^{-x-1} \sim A_1\alpha_2^{-x-1} \sim A_1(1 + \sqrt{2k/\sigma^2})^{-x-1}$$

and peculiarly†

$$\phi(0) = \phi_0 \sim A_1 \sim (1 - \phi_0)/4N\sigma\sqrt{2k},$$

$$\phi_0 \sim \frac{1}{1 + 4N\sigma\sqrt{2k}}.$$

---

* There may be other roots than α_1 of modulus < 1, but they have residues much smaller than A_1.

† For $x > 0$, we find $\phi(x) = \phi(-x)$; the coancestry is obviously the same for two opposite values, x and $-x$.

So we obtain the general formula (the sign $\sim$ meaning "only when k is small"):

$$\phi(x) \sim \frac{e^{-\sqrt{2k}\frac{x}{\sigma}}}{1 + 4N\sigma\sqrt{2k}}.$$

The numerical decrease with distance is the same as in the unidimensional continuous case; it may be seen that ϕ_0 also is the same; but in the two-dimensional case, ϕ_0 is very different (depending, as we have seen, on $\log 2k$ and not on $\sqrt{2k}$).

Bibliography

1. Allen, G. 1965. "Random and nonrandom inbreeding." *Eugenics Quarterly*, **12**:181–98.

2. Bernstein, F. 1930. "Fortgesetzte Untersuchungen aus der Theorie der Blutgruppen." *Z. für Abstamm. u. Vererbungslehre*, **56**:233–73.

3. Fisher, R. A. 1918. "The correlation between relatives on the supposition of Mendelian inheritance." *Trans. Royal Soc. Edinburgh*, **52**:399–433.

4. ———. 1930. *The Genetical Theory of Natural Selection*. Clarendon Press.

5. Galton, F. 1889. *Natural Inheritance*. Macmillan.

6. Haldane, J. B. S. 1939. "The spread of harmful autosomal recessive genes in human populations." *Annals Eugenics*, **9**:232–37.

7. Hogben, L. 1932. "The correlation of relatives on the supposition of sex linked transmission." *J. Genetics*, **26**:417–32.

8. ———. 1933. "The effect of consanguineous parentage upon metrical characters of the offspring." *Proc. Royal Soc. Edinburgh*, **52**:239–51.

9. Johannsen, W. L. 1909. *Elemente der exacten Erblichkeitslehre*. Jena: Fischer.

10. Lamotte, M. 1951. "Étude des populations naturelles de *Cepaea nemoralis*." *Bulletin Biologique de France et de Belgique*, supplément 35.

11. Malécot, G. 1939. *Théorie mathématique de l'hérédité mendelienne généralisée.* Paris: Guilhot, Traité des Sciences, 100 pp.

12. ———. 1944. "Sur un probleme de probabilities en chaine que pose la genetique." *Comp. Rend. Acad. de Sci.*, **219**:379–81.

13. ———. 1950. "Quelques schémas probabilistes sur la variabilité des populations naturelles." *Annales Univ. de Lyon, Sciences*, series A, **13**:36–60.

14. ———. 1951. "Un traitement stochastique des problèmes linéaires en génétique." *Annales Univ. de Lyon, Sciences, A.*, **14**:79–117.

15. ———. 1965. "Identical loci and relationship." *Proc. Fifth Berkeley Symposium on Mathematical Statistics and Probability*, IV, 317.

16. ———. 1966. *Probabilités et hérédité.* Presses Universitaires de France.

17. Pearson, K. 1904. "On a generalized theory of alternative inheritance with special reference to Mendel's laws." *Phil. Trans. Royal Soc.*, series A, **203**:53–86.

18. ———. 1909. "On the ancestral gametic correlations of a Mendelian population mating at random." *Proc. Royal Soc. London*, series B, **81**:255–29.

19. Philip, U. 1938. "Mating systems in wild populations of *Dermestes vulpinus* and *Mus musculus.*" *J. Genetics*, **36**:197–211.

20. Snow, E. C. 1911. "On the determination of the chief correlations between collaterals in the case of a simple Mendelian population mating at random." *Proc. Royal Soc. London*, series B, **83**:37–55.

21. Teissier, G. 1944. "Equilibre des genes lethaux xans les populations stationaires panmictique." *Rev. Scientif.* **82**:145–59.

22. Wright, S. 1931. "Evolution in Mendelian populations." *Genetics*, **16**:97–159.

23. ———. 1939. *Statistical genetics in relation to evolution.* Actualités Scientifiques et Industrielles, 802. Paris: Hermann.

24. ———. 1946. "Isolation by distance under diverse systems of mating." *Genetics*, **31**:39–59.

25. Weiss, G. H., and M. Kimura. 1965. "Steppingstone model of genetic correlation." *J. Applied Probability*, **2**:129.

Index

albinism, 7, 15
American Indians, 35–36
autosomes, 6

Bayes' formula, 64
Bernstein, F., 16
Bessel function, 73
blood groups, $7n$, 15, 35–36

Cepaea nemoralis, 75
coancestry, 23, 32, $70n$, $83n$
  coefficient of, 8–11, 16, 22, 32, 40f, $65n$, 66, $70n$, 73, 81
consanguinity, 14–16, 32, 43, 51, 53
correlation, 13–30
  coefficient of, 21–25, 27–29, 70
  a priori, $65n$, 66
  conditional, 43–44
  fundamental, 23, 25
  partial, 26
crossing over, 6, 17

Dermestes vulpinus, 14
disjunction, 5, 6, 21
dominance, 2–4, 19, 23–30, 57–58
  and mutation, 7

equilibrium, 17, 69
  asymptotic, 52, 63–64

evolution, 31–84
  and homogeneity, 35, 37
  and migration, 42
  and mutation, 7

Fisher, R. A., 23, 34
Fourier transform, 70f, 81
Fredholm iteration, 67f
Friedrich's ataxia, 15

Galton, F., 3f, 21f, 26
Gauss, K. F., 26, 62–63, 77
  law of probability of, 4, 21, 53f

Haldane, J. B. S., 34
Hardy's law, 14
Helmholtz equation, 73f, 77
hemophilia, 6
heredity
  continuous or blending, 3, 21
  discontinuous, 3
  multifactorial, 3f, 21
heterosomes, 6

inbreeding, coefficient of, 10, 16, 18f, 29, $75n$, $80n$
  and coefficient of coancestry, 9, 23, 43, 70

conditional, 53
and consanguinity, 14–16
isogamy, 13–30

Johansen, W. L., 21

Kimura, M., 74

Lamotte, M., 75
Laplace transformation, 55f
Liapounov's theorem, 21
linkage, 6
sex, 6, 29

Markov chain, 52, 65
theorem, 53, 56
mating
assortive, 14
random. *See* panmixia
Mendel, G., 20f, 75
laws of, 1–8 *passim*, 26, 30
migration, 39, 41, 48, 64–84
and selection, 42–43, 45f, 57–60
Mirabilis jalapa, 1
multiallelism, 7, 15, 20, 29
mutation, 21, 39–41, 65, 72
of blood groups, 35f
and evolution, 7

rate of, 11, 49, 51, 53, 57, 59, 76
and selection, 41–53 *passim*, 57, 59

panmixia, 14–21 *passim*, 31–37 *passim*,
44n, 67
in a finite, isolated population, 75–76
Pearson, K., 3f, 21f, 26
Philips, U., 14

selection, 31, 35, 41–64, 65n
coefficient of total, 45
gametic, 43, 45, 48–49, 51
zygotic, 44–46, 51–52
Snow, E. C., 22
stature, 3f, 7, 18

Taylor's formula, 58
development of, 68
Teissier, G., 46

variance, 20, 53ff, 59, 66, 74, 82
conditional, 53
dominance, 28
genic additive, 25, 28
total, 25, 28
of a trait, 20f

Weiss, G. H., 74
Wright, S., 57, 64, 76